AN INTRODUCTION TO
PHASE-INTEGRAL
METHODS

JOHN HEADING

DOVER PUBLICATIONS, INC.
MINEOLA, NEW YORK

Bibliographical Note

This Dover edition, first published in 2013, is an unabridged republication of the of the work originally published in 1962 by Methuen & Company Ltd, London, and John Wiley and Sons, Inc., New York.

International Standard Book Number

ISBN-13: 978-0-486-49742-6
ISBN-10: 0-486-49742-9

Manufactured in the United States by Courier Corporation
49742901
www.doverpublications.com

Contents

Preface   *page* vii

I   Historical Survey of the Problem   1

II   The W.K.B.J. Solutions   25

III The Stokes Phenomenon   48

IV   One Transition Point   69

V   Two Transition Points   94

VI   Applications to Physical Problems   116

     Appendix: The Relationship Between Series and
     W.K.B.J. Solutions   141

     Bibliography   152

     Index   157

Dedicated to my
two small boys
JEREMY AND PETER
who favoured me with the time
in which to write this book

Preface

He is a bold writer who puts his hand to the production of the first text (albeit a small one) devoted exclusively to the subject of phase-integral methods, since in such an exacting subject as mathematics it is surprising that the development of this technique over the last fifty years has been the occasion of so much error, criticism and dispute. Moreover, the treatment of the subject in the literature ranges from the ridiculously simple void of all rigour to the most sophisticated, the former hardly deserving mention and the latter not forming part of what is commonly known as the W.K.B.J. method. Hence the author has picked his way between the two extremes in order to produce what he hopes will prove to be a readable and helpful account of the method, since its name is known to many but its actual technique is known to but few. Moreover, since the number of applications is vast, only a brief selection of topics is considered, concentration being focused upon the method itself rather than upon detailed applications; it is sufficient to state that the author hopes to produce a more comprehensive mathematical account elsewhere at a later date.

The text has been developed to show that the process, once derived, is one of essential simplicity, being merely the application of certain well-defined rules, although an element of controversy still cannot be ruled out. To this end, the author has not hesitated to introduce his own notation consistently to denote the W.K.B.J. solutions, thereby to avoid the usual conglomeration of definite integrals within exponentials that habitually characterises the printed page dealing with phase-integral methods. Finally, the author wishes to thank the Editor's referee, Dr. K. G. Budden of the Cavendish Laboratory, Cambridge, for many helpful comments and corrections.

J. H.

University of Southampton
August 1961

CHAPTER I

Historical Survey of the Problem

1.1 Introductory remarks

A linear differential equation of the first order

$$w' + p(x) w = 0$$

(where a prime is used for differentiation with respect to the independent variable) possesses an immediate solution in the form

$$w = \exp\left[- \int p(x)\, dx \right].$$

On the other hand, an equation of the second order with variable coefficients such as

$$w'' + p(x) w' + q(x) w = 0$$

possesses no such simple general solution. Since many problems that arise in theoretical physics lead to differential equations of this type, attempts must be made to produce satisfactory methods of solution.

If p and q are specially chosen, the various standard transcendental differential equations are produced. For example, if $p = 1/x$ and $q = 1 - v^2/x^2$, we obtain *Bessel's equation* of order v; if $p = 0$ and $q = -x$, we obtain the *Airy equation*. These special cases are singled out because exact solutions are readily obtainable in terms of contour integrals, from which the *power-series solutions* in ascending powers of x and the *asymptotic solutions* in descending powers of large x may be obtained.

In all but standard cases, contour-integral solutions can no longer be found, and similarly the explicit forms of the coefficients occurring in the power-series and asymptotic solutions cannot be written down. Under these circumstances, approximations are necessary, and *phase-integral methods* or the W.K.B.J. solutions constitute a powerful technique for dealing with these approximations. Bluntly speaking, the technique is to restrict oneself to the first term of the asymptotic series; the extraordinary difficulties that arise by using

1

such a simple approximation must be investigated and smoothed out by means of a series of rules by which such approximations may be manipulated without fear of mathematical error.

The use of approximations in mathematical physics is so familiar that it is surprising that some have complained about the point of view adopted in phase-integral methods. For example, with no just foundation for such remarks, Smyth [102] has criticized a paper making use of the method by writing: 'It should be observed that the authors have used a solution which is a very poor approximation to the given problem as an approximate solution to another problem. It is certainly not to be expected that the results obtained in this manner will have any connection with the original problem.' Concerning approximations, Schelkunoff [97] has more wisely remarked that there is 'something in human nature that makes one yearn for the exact answer to a given problem. In particular it makes little difference whether a given problem is solved approximately or replaced by an approximating problem which is then solved exactly.' Introducing a new approach that leads to a difference in applicability, Hines [59] has observed that his new method yields 'an approximate evaluation of the exact solution, rather than an exact evaluation of an approximate solution as is found in the W.K.B.J. method'.

Phase-integral methods are applied to the particular equation

$$w'' + h^2 q(x, h) w = 0, \tag{1.1}$$

in which h is a large parameter and $q(x, h)$ a variable function of x and h. The real independent variable x may often be replaced by the complex variable z. Under certain circumstances it will be shown that approximate solutions of this equation are

$$w = q^{-1/4} \exp\left(\pm ih \int^z q^{1/2} dz \right) \left[1 + O\left(\frac{1}{h} \right) \right], \tag{1.2}$$

the error term of the form shown being maintained uniformly throughout certain *restricted* domains of the complex z-plane. A zero of q in the complex plane is called a *transition point*; evidently the expressions (1.2) cease to be valid solutions near a transition point owing to the singularity caused by the factor $q^{-1/4}$, while the exact solution of equation (1.1) must obviously remain finite at such a point. These expressions are known as the *phase-integral solutions* or the W.K.B.J. solutions, after Wentzel [115] (1926), Kramers [72] (1926), Brillouin [21] (1926) and Jeffreys [64] (1923). Concerning the common name given to these solutions, Bailey [6] has chosen to call them 'the L.R. approximations' after Liouville [79] (1837) and Rayleigh [94] (1912). Moreover, Bailey has written: 'The custom, based on historical ignorance, of using the titles "B.K.W." or "W.K.B." (or some other permutation of these three letters) is wrong as it does such flagrant injustice to the truth.' Jeffreys [66] has often called these solutions 'approximations of Green's type', after Green [48] (1837).

The object of phase-integral methods is to show how such solutions may be traced about the complex z-plane beyond their original restricted domains of validity.

In connection with solution (1.2), it should be stressed that we may discuss errors between an approximate solution and an exact solution even when there exists no exact *analytical* solution of the differential equation. An exact solution exists, not because some analytical form can necessarily be found for it, but because what are known to mathematicians as *existence theorems* demonstrate the existence of such a solution; see, for example, Chapter XII of the text by Ince [127], *Ordinary Differential Equations*. The error term $O(1/h)$ occurring in solution (1.2) implies that an exact solution would contain a term $F(z, h)/h$, where $|F(z, h)|$ is bounded by the inequality $|F(z, h)| < M$, for all z in the domain under consideration and with h greater than some value h_0. There is no implication that h must tend to infinity in (1.2); all we assert is that h must be large. On the other hand, z may or may not tend to infinity in the domain under consideration. Hence problems that require finite values of h and z are embraced by solution (1.2).

Historically, the use of such approximate solutions may be traced to Carlini (1817), whose work may be referred to in Watson [114] (page 6). Carlini considered a specific equation, being in effect Bessel's equation, for which he derived an approximate solution valid when n is large and for $0 < x < n$. Liouville [79] (1837) and Green [48] (1837) produced asymptotic solutions for more general equations, but void of rigour, for a range of x in which no transition point occurs. Green was concerned with the propagation of long waves in a channel of non-uniform section, provided that the period is short enough for the depth and width of the channel not to vary greatly within a wavelength; he was thereby able to show that for tidal waves in such a channel energy is transmitted without reflection losses.

The name of Horn [60] (1899) should also be mentioned in this connection, though strictly speaking his name belongs to the particular investigators detailed in section 1.5. He is usually stated to be the original principal contributor to the theory of the existence of asymptotic solutions in intervals free from transition points. From the W.K.B.J. point of view, we may observe that Jeffreys [64] has remarked that Horn did not give the most convenient form for the first term of the solution.

1.2 The Stokes phenomenon

Sir George Gabriel Stokes (1819–1903) would have been further honoured had it been realized then that without the existence of the discontinuous changes in the arbitrary constants that occur in the asymptotic solutions of certain differential equations, the reflection of waves by inhomogeneous media (such as the ionosphere) would have been an impossibility. Stokes had been confronted by this phenomenon in his study of Bessel functions, and evidently was troubled with the difficulty for some years before finally getting to the root of the matter. On March 19th, 1857, Stokes [103] wrote to his young lady: 'I have been doing what I guess you won't let me do when we are married, sitting up till 3 o'clock in the morning fighting hard against a mathematical difficulty. Some years ago I attacked an integral of Airy's, and after a severe trial reduced it to a readily calculable form. But there was one difficulty about it which, though I tried till I almost made myself ill, I could not get over, and at last I had to give it up and profess myself unable to master it. I took it up again a few days ago, and after two or three days' fight, the last of which I sat up till 3, I at last mastered it.'

In his first paper on the subject, Stokes [104] (1857) considered the equation

$$w'' - 9zw = 0$$

in the complex z-plane. (Nowadays, the factor 9 would be omitted.)

He gave two independent power-series solutions of this equation, which for small $|z|$ could be used in the computation of the general solution, which would involve two *fixed* arbitrary constants multiplying the two power series respectively. Secondly, he gave the two independent asymptotic expansions of the equation (the leading terms being proportional to our expression (1.2) with $q = -9z$ and $h = 1$). He noticed that if for a certain range of $\arg z$ a general solution was represented by a certain linear combination of the two asymptotic solutions, then in a neighbouring range of $\arg z$ it was by no means necessary for the same linear combination of the two fundamental asymptotic expansions to represent the same general solution. In fact, Stokes found that the constants of the linear combination changed discontinuously as certain lines (given by $\arg z =$ constant) were crossed; today, these are called *Stokes lines*. Finally, by means of an integral representation of the solution over the whole complex z-plane, he showed how the two fixed constants of the series solution were associated with the corresponding (changing) constants of the asymptotic solution. In the Appendix, we show by a simple method (partially following Stokes [106] (1889)) how these constants are linked together. The Stokes phenomenon and the Airy integral is discussed in Chapter III, while an account based on contour integration may be found in Budden's text [24], Chapter XV.

Briefly, the reason why a discontinuity is allowed in a solution that is strictly continuous is because asymptotic series *always contain an inherent error*, and the discontinuity that occurs is less than this inherent error. An asymptotic series is really divergent, but nevertheless it can be summed with great accuracy for large $|z|$ by employing only those terms for which the magnitude of the terms decreases along the series. The order of magnitude of the error at any stage is equal to the magnitude of the first term omitted. Maximum accuracy is therefore attained by summing up to the term of smallest magnitude, after which the magnitude of the terms increases without limit yielding a series that would ultimately diverge (see Whittaker and Watson [116], Chapter VIII).

In a second paper, Stokes [105] (1871) considered the constants for the asymptotic expansions of the solutions of Bessel's equation; the relevant results are derived in our section 3.6.

Throughout our text, these changes in the constants are effected by simple rules based upon factors known as *Stokes constants*. Here, we are only concerned with these constants for the Airy equation, Bessel's equation and Weber's equation. Their values for Whittaker's confluent hypergeometric function have been given recently by Heading [55], who also considered their values in a series of papers [51, 52, 53] for the nth order equation

$$w^{(n)} = (-1)^n z^m w.$$

It should be mentioned that asymptotic expansions may now be summed by more powerful methods based on *converging factors*. This technique may be examined in a series of papers by Dingle [26, 27]; the Stokes phenomenon is still operative, however, in this modern method.

1.3 The W.K.B.J. approximations

The early use of these approximations was spasmodic until their applicability was popularized by the authors whose initials they bear. Lord Rayleigh [94] (1912), concerned with the propagation of waves through a stratified medium, actually derived an approximate solution using a cosine instead of an exponential as in our equation (1.2). His solution of the equation

$$\phi'' + \kappa^2(x)\phi = 0 \qquad [\kappa(0) = 0]$$

was expressed in the form

$$\phi = \kappa^{-1/2} \cos\left(pt - \int \kappa \, dx\right).$$

Discussing waves on a membrane, he actually replaced $\kappa^2(x)$ by a linear approximation through the transition point $x = 0$ where his solution broke down, and he observed that total reflection occurred for a solution that became vanishingly small in the range for which $\kappa^2(x)$ was negative. He did not however continue the investigation and thereby lost the chance of being the originator of the technique for dealing with transition points.

Gans [41] (1915), whose work is discussed in section 1.6, considered the propagation of light in a slowly-varying stratified medium; he appears to be the first investigator to examine the effects of a transition point in a systematic way.

Fowler *et al.* [35] (1920), concerned with the aerodynamics of a spinning shell, were confronted with two simultaneous differential equations, totally of the third order. They followed Horn's [60] method, and actually produced the standard W.K.B.J. solutions for their equation $y'' + My = 0$.

Concerned with the application of Mathieu's equation to the free oscillations of water in an elliptical lake, Jeffreys [64] (1923), though chiefly interested in the effect of a transition point on the solutions, was the first to systematize the W.K.B.J. solutions. As in his following papers on the same subject, he considered an equation of the form

$$w'' = (h^2 \chi_0 + h\chi_1 + \chi_2) w \equiv h^2 q w,$$

say, where χ_0, χ_1, χ_2 are real functions of x along the real axis, and h is a large parameter. Jeffreys has recommended taking for the integral occurring in the solutions (1.2)

$$\int \chi_0^{1/2} \left(1 + \frac{\chi_1}{2h\chi_0}\right) dx,$$

rather than $\int q^{1/2} dx$. He has often quoted an exceptional case (see Jeffreys [66], section 3.3) in which the use of his recommendation is essential rather than the use of the more common $\int q^{1/2} dx$.

Throughout our text, we shall use $\int q^{1/2} dx$, thereby restricting ourselves to those functions q for which such a form is valid; a condition given in section 2.4 must in fact be satisfied. In physical problems it seems likely that this condition is satisfied, and that Jeffreys' exceptional case is of mathematical interest only.

The errors associated with the use of the W.K.B.J. solutions are always expressed as $O(1/h)$ times the main approximation, as shown in expression (1.2). This is of necessity vague, but satisfactory. Olver [90] (1962) has investigated the error bounds for the W.K.B.J. solutions (or the Liouville-Green approximations, as he calls them), and has applied the theory to parabolic cylinder functions and modified Bessel functions of large order.

1.4 Asymptotic uniformity in the complex plane

When the independent variable is complex, and when a particular W.K.B.J. solution is examined, say the one in which the sign is positive, the error for a particular domain in the z-plane is $O(1/h)$ times this solution. If now z is moved about the plane, this particular W.K.B.J. solution remains satisfactory provided the error remains of the same form. However, it is found that z may vary into a domain in which the error ceases to be of this form, but changes to $O(1/h)$ times the *other* W.K.B.J. solution (with the minus sign). When this happens the new form of the error is exponentially large compared with the magnitude of the original W.K.B.J. solution. In other words, we have traced the solution into a domain in which the original approximate representation is no longer valid. Boundaries across which such a change takes place are known as *anti-Stokes lines*; within such boundaries, the solution is *uniformly asymptotic*.

Jeffreys [66] (1953) examined this situation for one transition point using both the W.K.B.J. approximations and the more complete approximations based on the Airy functions (see our section 1.11). However, he did not elucidate the question of the boundaries in a systematic way. Olver [89] (1959) examined the equation with two transition points whose exact solutions are expressed in terms of Weber functions. Using both the W.K.B.J. approximations and the approximations based on the Airy integral function, he considered the boundaries necessary for the maintenance of uniformity. Heading [56] (1962) has considered the case of an arbitrary number of transition points, examining the boundaries of regions of uniform validity both for the W.K.B.J. solutions and for more general approximations.

Many writers treat the subject of phase-integral methods without concerning themselves with this question, but this stands as an inherent weakness of their treatment. For example, Budden [24] in his comprehensive account of the subject in his text, *Radio Waves in the Ionosphere*, realizes this weakness but nevertheless follows the majority of writers and thereby fails to standardize a large parameter in terms of which parameter the uniformity of the approximations may be examined, and reviewers [57, 108] have called attention to this fact.

1.5 Asymptotic series

Horn's [60] paper formed the basis for mathematical investigation into the actual asymptotic series associated with solutions of differential equations containing a large parameter. Not only have single differential equations been considered, but also systems of simultaneous linear differential equations expressed in matrix form. The classical case of the domain free from transition points has long since been investigated, and later writing has focused upon equations possessing transition points. The names of Schlesinger [98] (1907), Birkhoff [11] (1908), Tamarkin [107] (1928) and Turrittin [110] (1936) deserve mention prior to the more modern voluminous writings of Langer, Olver and their contemporaries. A collection of references to this recent work may be found in Olver [89] (1959). It cannot be said, however, that this work forms part of what is commonly understood by phase-integral methods, although some aspects of it were developed for physical problems and other aspects for purely numerical purposes.

1.6 Connection formulae

Consider equation (1.1) in which $q(x)$ is real for real x, such that $q(x) = 0$ and $q'(x) \neq 0$ when $x = 0$. Suppose that also $q(x)$ is negative when x is positive, although other writers may well reverse the whole discussion. Evidently from (1.2), the W.K.B.J. solutions contain an exponential of a real quantity for $x > 0$, one of which is exponentially large for large real h and the other exponentially small (these will be called *dominant* and *subdominant* respectively). For $x < 0$, the solutions contain an exponential of a purely imaginary quantity, and may represent progressive waves physically; sines and cosines of real argument may in fact be used instead.

The problem to be investigated is this: Given a linear combination of the two solutions for $x < 0$, it is apparent that such a representation of the solution cannot be extended up to $x = 0$. What will be the proper form for the solution for $x > 0$? *Connection formulae* provide the answer to this question. A substantial *metamorphosis* evidently takes place in the form of the function, under which sines and cosines must smoothly join on to exponential functions.

This problem is treated with various degrees of rigour and clarity in several texts. For example, the reader may consult Jeffreys and

Jeffreys [68], *Methods of Mathematical Physics*, or Morse and Feshbach [83], *Methods of Theoretical Physics*, or various books on wave mechanics, such as Kemble [70], *The Fundamental Principles of Quantum Mechanics*, or Bohm [13], *Quantum Theory*.

Lord Rayleigh [94] (1912) touched the fringe of the subject in considering the propagation of waves on a membrane, but Gans [41] (1915), concerned with the propagation of light in a slowly varying stratified medium, appears to have been the first to treat the matter seriously. Replacing the actual function q by a linear approximation through $x = 0$, he solved the resulting equation exactly in terms of Bessel functions of order $\frac{1}{3}$ (the Airy integral had not been standardized then). Providing the medium was sufficiently slowly-varying, the asymptotic expression for the Bessel function could be used before the linear approximation broke down seriously. These asymptotic expressions were then connected directly to the W.K.B.J. solutions valid on each side of the transition point. Owing to the Stokes phenomenon, different combinations of the two W.K.B.J. solutions had to be used on either side of the zero. Such a procedure as this always implied total reflection, seemingly taking place at $x = 0$. His method was equivalent to 'patching' three distinct approximate solutions together, in the three overlapping ranges $x < 0$, $x \doteq 0$, $x > 0$. The theory was not, however, expressed in a form easily adaptable to a general function $q(x)$.

The first systematic connection formulae were produced by Jeffreys [64] (1923), whose investigations were carried out without the knowledge of Gans' previous paper. Approximating to the profile through the transition point by a linear gradient, and identifying the coefficients of the asymptotic expansions of the known solutions of the equation with the modified profile and the coefficients of the W.K.B.J. solutions of the given equation in ranges of x where overlapping validity occurred, he obtained two connection formulae in the form

$$\text{trigonometric solution} \leftrightarrow \text{subdominant solution}, \qquad (1.3)$$

$$\text{trigonometric solution} \leftrightarrow \text{dominant solution}. \qquad (1.4)$$

He defined the symbol $\leftrightarrow$ by stating that it 'indicates that the functions it connects are asymptotic approximations to the same function under different circumstances'.

A trivial error in the phase constant occurring in the trigonometric expression on the left-hand side of (1.4) has been noted by Langer [74]. Langer has also pointed out the danger of the use of the symbol $\leftrightarrow$, showing that it can give rise to 'misconceptions that have not failed to show themselves in the literature'. Several authors have supposed that the symbol permits a W.K.B.J. solution to be traced from the left to the right or from the right to the left, namely that the left implies the right and the right implies the left. This misconception arises because the definition of the symbol given by Jeffreys was not understood. It would perhaps have been more appropriate to use the symbol $\leftarrow w \rightarrow$, implying that there is a solution w, possessing the stated asymptotic expressions on the left and on the right; w necessarily implies the W.K.B.J. solutions but not *vice versa of necessity*.

Jeffreys applied his formulae to Bessel's and Mathieu's equation.

On the right-hand side of (1.4), Jeffreys had the exponential e^{-M}, where M was the phase-integral and was actually negative, thereby yielding an exponentially large dominant solution. These connection formulae were used by Mott [84] (1928) to derive the laws of classical statistical mechanics as approximations to those of quantum mechanics. Unfortunately, he misquoted Jeffreys' paper, overlooking the fact that M was negative. Jeffreys has observed that this mistake was repeated in the first edition of Fowler's *Statistical Mechanics* and apparently attributed to him in the second edition. The correct version was given by Mott and Massey [85] (1933).

Later, Jeffreys [65] (1942) returned to the subject of his connection formulae, and re-derived them by means of the standardized solutions of the Airy equation. The previous symbol M is now changed in sign to avoid further confusion, but what is more important, he produced an entirely new version of the dominant connection formula (1.4); (the actual forms are considered in our section 4.5). The error in the right-hand side of (1.4) is always greater than the subdominant solution (1.3), so an arbitrary multiple of (1.3) may be added to (1.4) at will. The new left-hand side of (1.4) will then be completely changed, but the right-hand side will still appear merely with a dominant expression, the subdominant term being dropped. Now Jeffreys' second version of (1.4) occurs quite naturally from the asymptotic expressions quoted for the standardized Bi(x) function, and from that time onwards the older version has been dropped while the new version, stated to be 'more convenient' has been habitually used.

Moreover, this new version has often been used in the sense that the right-hand side implies the left. B. Jeffreys [63] (1942), for example, took this point of view in a potential barrier problem. Had she used the original version of (1.4) she would have obtained a different answer.

In a more recent paper, Jeffreys [67] (1956) continues to argue against Langer's suggestion that $\leftrightarrow$ should be replaced by $\rightarrow$ in the modified version of (1.4), this suggestion rightly indicating that the left-hand side *does* imply the right-hand side. He goes on to say that if such were so, no information could be deduced in problems with boundary conditions on both sides of the origin, and 'in some cases results have had an accuracy far greater than there was any apparent reason to expect'.

Jeffreys then considers two problems. In the first, a perfect reflector exists at $x = b > 0$ and at $x = -a < 0$, yielding an eigenvalue problem. An answer is obtained by a method which tacitly uses the modified version of (1.4) from right to left, though Jeffreys strenuously denies that such is the case. In point of fact, and Jeffreys does not seem to have appreciated this, the same answer would be obtained had the first version (1.4) been used, or any other version obtained from it by adding an arbitrary multiple of (1.3); the reason for this may ultimately be traced to the fact that there is no energy transmission associated with a standing wave pattern.

Secondly he considers the problem of the potential barrier, again in effect using the modified version of (1.4) backwards, though this is not explicitly evident in his calculation. The answer obtained is correct when compared with the known exact answer found by other methods for a parabolic barrier, for which he quotes the present author's thesis [50]. Moreover, it should be stressed that the modified version of (1.4) is now the *only* one that yields the correct answer; all others would be wrong. Jeffreys, then, still justifies his second version on the grounds that it yields correct results. The fundamental reason why this yields correct results when used from right to left is explained in section 4.9, and the author believes this explanation is original. It concerns problems in which $q(x)$ is real for real x, and for which, physically speaking, energy flow is conserved along the x-axis.

After Jeffreys' original paper, the need for solutions of Schröd-inger's equation in wave mechanics had priority. Wentzel's [115] (1926) contribution should be mentioned, while it is due to Kramers [72] (1926) that the 'so-called Stokes phenomenon' came to be associated with the subject, though by name alone. Both investigators were really concerned with eigenvalue problems and potential wells. Goldstein [46,47] (1928, 1931) considered connection formulae for the W.K.B.J. solutions when q had a zero of multiplicity greater than unity (see our section 4.10), while Zwaan [118] (1929) appears to have been the first to use the complex z-plane to investigate transition points, even though $q(x)$ is still real along the real axis. A summary of his thesis may be conveniently found in one of Langer's [74] papers. He traced a subdominant W.K.B.J. solution from one side of the transition point to the other by tracing it round the point in the complex plane. When the real axis was again reached, the condition that the solution must be real provided sufficient information for the deduction of the appropriate connection formula. Langer has pointed out the weaknesses of such a treatment. Kemble [70], in his text, employed the same ideas, but more rigorously.

Furry [37] (1947) appears to have been the first author to have treated the idea of the Stokes phenomenon seriously. He sought to break away from Eckersley's approach (see our section 1.9), his criticism being that 'no readily understandable and convincing argument for its validity has ever been published'. In his paper, a transition point is marked as a point in the complex z-plane, with three Stokes lines and three anti-Stokes lines radiating from it. He calculated the change that the subdominant coefficient must undergo on the Stokes lines, from which he immediately deduced Jeffreys' connection formulae. His general rule was, however, very ill-expressed. It may be remarked that this concept, though not the method nor the definitions, forms the basis of the theoretical chapters of this text.

He has been followed in detail in Kerr [71] (1951), *Propagation of*

Short Radio Waves, section 2.8, by Freehafer, though many parts of the theory are left untouched. Concerning Eckersley's work, Freehafer remarked: 'The validity of the phase-integral method has not been established in general. Each application must stand on its own merits. In particular, the method is based on the tacit assumption that z_0 is an isolated zero', z_0 being the transition point.

1.7 The potential well

If q is such that q is real for real x, and possessing two real zeros at x_1 and x_2 with q positive between them, we have the problem of the *potential well*. If $q = a^2 - x^2$, the equation may be solved exactly in terms of parabolic cylinder functions. Boundary conditions are assumed such that $w \to 0$ as $x \to \pm \infty$; this leads to an eigenvalue problem, the discrete values of the total energy being required in problems using Schrödinger's equation.

Jeffreys' connection formula (1.3) is used for both transition points x_1, x_2, and this leads to the formula

$$\exp\left(2ih \int_{x_1}^{x_2} q^{1/2} dx\right) = -1,$$

or

$$2h \int_{x_1}^{x_2} q^{1/2} dx = (2n+1)\pi, \tag{1.5}$$

n being an integer. Dunham [28] (1932) used Zwaan's method for deriving this result. But this theory assumes that the W.K.B.J. solutions are valid between the zeros. If this is not so, this equation still gives the correct answer for the true harmonic oscillator, since exact solutions may be obtained in this particular case. Slater [100], in his text, has called this 'an interesting fact', but few have used phase-integral methods alone to prove that the correct answer will in fact be produced in this case. Birkhoff [12] (1933) has examined this special case, and a more simple treatment is found in our section 5.5.

Eckart [29] (1960) has recently produced the equivalent of equation (1.5) in his text, *Hydrodynamics of Oceans and Atmospheres.* Observing that 'in recent years there has been a revived interest in the W.K.B.J.

approximation, and noteworthy advances have been made', he derived equation (1.5) 'by very simple ideas', merely by considering the number of nodes in the range x_1 to x_2. He considered the W.K.B.J. solution as representing a wave reflected at x_1 and x_2, and to ensure a whole number of wave forms he wrote immediately

$$h \int_{x_1}^{x_2} q^{1/2} dx = (n+\Delta)\pi,$$

where Δ remains undetermined, representing a sort of phase shift upon reflection. Such a 'simple idea', can, alas, have no foundation at all, since the W.K.B.J. solutions are not valid near x_1 and x_2, so the idea of counting nodes falls to the ground. Eckart then proceeded to use the W.K.B.J. approximations (though without the $q^{-1/4}$ factor) in problems occurring in the investigation of the thermocline and the thermosphere, as well as in ray theory.

Bell [7] (1944) considered the eigenvalues for the total energy when the potential energy function has the form $|x^m|$, and he presented a numerical comparison between the approximate results for $|x|$ and the exact solution. Titchmarsh [109] has also considered the case of x^4 in his text.

Auluck and Kothari [4] (1945) have considered the case of the bounded harmonic oscillator, that is, an oscillator with perfectly reflecting walls outside the two transition points. They did not use the W.K.B.J. approximations, since the exact form of the solution was available. But in quoting the asymptotic expansions from Whittaker and Watson [116], they failed to realize that they required the value of the coefficient of a subdominant term in the presence of a dominant term. Special care is necessary to decide the coefficient to be used, and they were in error by a factor $\frac{1}{2}$; this problem is considered in our section 6.3.

Furry [37] (1947) examined the normalization of the W.K.B.J. solution for the approximate harmonic oscillator, using a method that improved the technique of earlier writers (see our section 5.6).

The hydrogen atom has fared badly under early writers owing to the singularity that occurs at $r = 0$ in the functional form of q for the radial wave function; this arises from the electrostatic potential energy $-e^2/r$. Langer [75] (1937) has quoted six authors who had to use the known exact solution in order to adjust the phase-integral method to yield its best answer. This does not speak well of phase-integral methods if such external props had to be used. For example, Young and Uhlenbeck [117] (1930) merely use equation (1.5) based on the radial wave equation in r as it stands, expecting a subdominant solution within the lower transition point to remain finite at the singularity $r = 0$, but such is not the case. The results they derived are, then, correct only when l is large, where l is related to the separation constant introduced when the r and θ equations are separated.

Langer [75] (1937) has given the corect calculation, showing that a preliminary transformation is necessary to deal with the singularity at $r = 0$. Phase-integral methods then yield the correct values of all the energy levels (see our section 6.4). The full approximate forms of the corresponding wave functions may be found in Landau and Lifshitz [73], *Quantum Mechanics*.

1.8 The potential barrier

Let $q(x)$ be real for real x, but possessing zeros at x_1 and x_2 such that $q(x) < 0$ between them. Hartree [49] (1931) mentioned this case in considering isotropic ionospheric radio propagation. Stating that Jeffreys' (1923) method was not applicable when q had two zeros, he observed that if they were widely spaced a similar method would be applicable across each zero separately, since the W.K.B.J. solutions could be used between them. Hartree did not, however, attempt this mathematically; indeed, wrong results would have been obtained had he used Jeffreys' first version (1.4).

Kemble [70] (1937) gave a partial treatment in his book, although one would not suppose that the treatment was only partial by glancing at the mathematical complications of his text. Gamow [38] (1928) considered a square potential barrier for which of course exact solutions can easily be found; he used the boundary conditions that w and w' are continuous across the discontinuities. But later, Gamow [39] (1937) produced a similar treatment for a continuous potential barrier using the W.K.B.J. solutions in his text, *Atomic Nuclei and Nuclear Transformations*. Of this, B. Jeffreys [63] (1942) has written that it is the 'best known account, but open to very serious criticism, since the asymptotic solutions are used at points where they become infinite'. In a word, all Gamow did was to produce his own connection formula, by stating that the W.K.B.J. solutions and their derivatives were continuous over the transition points. Such a treatment must be void of all justification, yet it leads to the correct result in a certain special case, while other results are wrong by simple numerical factors. The later third edition [40] of the text merely reproduces these errors, although they had by then been pointed out by B. Jeffreys.

B. Jeffreys [63] (1942) used Jeffreys' connection formulae across the two transition points, obtaining the correct answer to a certain degree of accuracy. However, she used Jeffreys' modified version of formula (1.4) from right to left, and the modified formula, as we have already pointed out, is the only one that yields the correct answer. It appears, then, that it was a coincidence that the correct answer was obtained, since no explanation had then been given why that formula and none other was essential in such problems. Rydbeck [95] (1948) also gave an independent examination of the same problem. Several texts have followed B. Jeffreys' treatment; for example, Mott and Sneddon [86], *Wave Mechanics and its Applications*.

1.9 Applications of the method

Hartree [49] (1931) was the first investigator to apply these techniques to an isotropic loss-free ionosphere, considering two models with a linear and parabolic distribution of free electron density respectively. He solved the two problems firstly by ray methods, thinking in terms of the standard optical approach; then he solved the two problems by exact analytical methods, finding the asymptotic form of his answer for higher frequencies. He observed that the phase calculated by ray methods was $\frac{1}{2}\pi$ in error for both models compared with the exact asymptotic forms. Next, he treated both models by using Jeffreys' connection formula (1.3), and obtained results identical with those produced by ray methods, but with the additional term $\frac{1}{2}\pi$ included. It appears, then, that Hartree was the first to show that Jeffreys' method was equivalent to standard ray treatment plus correcting phase factors.

Eckersley [30] (1931) next entered the field, and in all his following papers persisted in his outlook. In a paper entitled 'On the connection between the ray theory of electric waves and dynamics', he introduced something new into propagation theory. Unfortunately he gave no proof, merely being guided by the success of certain hypotheses in quantum theory. Planck's original hypothesis for the motion of a particle moving with classical momentum p may be expressed in the form

$$\oint p \, dx = nh, \tag{1.6}$$

where n is an integer and h is Planck's constant. Schrödinger's wave equation also yields this result when the W.K.B.J. solutions are traced from one transition point x_1 to the other x_2 and back again; the single-valued nature of the solution then yields (1.6) if no phase changes upon reflection at x_1 and x_2 are introduced. Eckersley took this suggestion up, and applied it to the propagation of radio waves between the earth and the ionosphere. He stated that the 'relation of this method to the accurate calculation of the proper values is the same as the relation of the older Bohr-Sommerfeld phase-integral quantum method to the new Schrödinger wave mechanical method'. In a following paper, Eckersley [31] (1932) allowed a phase change to take place upon reflection by the ionosphere.

Later, Eckersley and Millington [33] (1938) considered the application of phase-integral methods to the diffraction of wireless waves round a spherical earth; that is, they calculated the various modes permitted in the field of an elevated dipole aerial and receiver above the earth's surface. Certain phase constants upon reflection were introduced by comparison with the exact solution given previously by Watson, and the authors rightly claimed that their methods showed up the physical significance of the mathematical complications of Watson's method. Had Jeffreys' connection formula (1.3) been used, the authors would have been able to deduce the phase constant without reference to the exact solution.

Following the publication of Jeffreys' connection formulae, many authors employed them in problems in quantum mechanics, but nothing essentially new was added to the theory. For example, Arnot and Baines [2] (1934) used both Jeffreys' approximations and the Born approximation in the calculation of approximate phases that occur in the theory of electron scattering.

Booker and Walkinshaw [15] (1946) in a paper entitled 'Mode theory of tropospheric refraction', encountered an equation whose solution could be found by phase-integral methods. Although this was a mathematical paper, progress was made by physical reasoning. They adopted approximate solutions of the form (1.2), and the occurrence of the factor $q^{-1/4}$ was justified on the grounds that it was required 'in order to make the mean vertical flow of energy independent of height, and so conserve energy'. Reflection at the transition point was explained according to transmission line theory; their 'impedance' was purely inductive above the transition point and infinite at the point itself. Their interpretation was that at the transition point there was a terminating inductance, associated with a storage field above. They observed: 'A phase advance of $\frac{1}{2}\pi$ on reflection always occurs when a transmission system of characteristic impedance Z is terminated by an inductive reactance iZ.' By this argument, they introduced the usual phase change $\frac{1}{2}\pi$, which can be justified by using Jeffreys' formula (1.3).

The extensive work of Bremmer relating to isotropic and anisotropic ionospheres may be found in three papers [16, 17, 18] (1949) and in his text [19], *Terrestrial Radio Waves*. Friedman [36] (1951) considered ducts that occur in the theory of propagation of electromagnetic waves in a non-homogeneous atmosphere, while Abbott [1] (1956), concerned with the propagation of bores in channels and rivers, used integrals suggested by the W.K.B.J. method. Seckler and Keller [99] (1959) considered acoustical diffraction in inhomogeneous media, using the phase-integral method to obtain the eigenvalues of a certain propagation constant, while Wasow [113] (1960) dealt with a problem in the adiabatic theory of Hamiltonian systems, obtaining two simultaneous linear differential equations. Finally, a translation of Gol'dman *et al.* [45], *Problems in Quantum Mechanics*, contains numerous interesting examples using what is termed the 'semi-classical approximation', while the apparent lack of any major Russian contribution is suggested by an examination of the excellent text, *Waves in Layered Media*, by Brekhovskikh [124].

1.10 Higher approximations

The W.K.B.J. solutions of the differential equation (1.1) are merely first approximations, and Bremmer [16, 20] (1949, 1951) has often stressed the physical interpretation to be placed upon further correction terms. He imagined a range of x free from transition points subdivided into a large number of thin homogeneous slabs (see our section 6.5), and the transmission of a wave is then accompanied by an infinite series of reflections backwards and forwards between the interfaces. The W.K.B.J. solution itself is the wave that arises when all secondary waves diverging from it by means of one reflection are discarded. Further approximations are obtained by taking into account the contributions from waves that have suffered one, two or more reflections. Bellman and Kalaba [8, 9] (1958, 1959) and Atkinson [3] (1960) have examined the convergence of the series thus obtained, and have shown that the limiting form as the individual slabs tend to zero thickness actually satisfies the original equation (1.1).

1.11 Approximations valid through the transition point

In all our previous considerations, transition points have been excluded from the domain of validity of the approximate solutions (1.2). Moreover, Jeffreys' connection formulae are associated with the ranges $x < 0$ and $x > 0$, while a special solution is used through $x = 0$ to join the two sides together. Now although this technique has been placed on a rigorous foundation, it proves to be unsatisfactory from some points of view. Langer [74] (1934), in a published lecture, complained that Jeffreys' method merely patched 'the results together into a more or less complete representation'. He therefore attacked the problem from a more fundamental standpoint, namely 'that of obtaining at one and the same stroke a representation which is valid over the whole domain of the variable'. In terms of modern functions, Langer was able to obtain an approximate solution to (1.1) involving the Airy integral valid at the transition point, and such that the asymptotic expressions of the solution were merely the W.K.B.J. solutions in domains away from the transition point (see our equation (2.10), and example 2, section 2.7). Wasow [113] (1960) has recently discussed the difference between Jeffreys' and Langer's approach.

This overall representation has been the main theme of modern writers dealing with asymptotic series (see section 1.5). Applications to physical problems have been less varied, since in the calculation of eigenvalues and reflection coefficients the value of the approximate solution *at* the transition point is not required. Langer [76] (1950) has used such solutions in considering microwave propagation in an atmosphere whose refractive index varies with height. Benney [10] (1961), in discussing the non-linear theory for oscillations in a parallel flow, derived an equation similar to (1.1) but with a function of the independent variable on the right-hand side. The method of variation of parameters was used once suitable approximate solutions for the left-hand side had been obtained. Following Langer, he introduced Hankel functions of order $\frac{1}{3}$ rather than the standardized Airy functions, thereby in effect introducing singularities where strictly speaking none really can exist. Takahashi [122, 123] has used such approximations in considering the dispersion of Love waves and Rayleigh waves in seismological theory, while Budden [125] (1961) has employed both the W.K.B.J. solutions and Langer's solution in discussing the reception of radio signals from outer space by an aerial situated in the upper part of the ionosphere.

Jeffreys [66] (1953) examined the representation of the solution through one transition point in terms of the Airy integral, but did not examine the boundary of the domain of validity in a systematic manner, and what happened at the transition point was not discussed. Heading [56] (1962) considered the case of an arbitrary number of transition points, expressing the boundaries in terms of anti-Stokes lines. Olver [89] (1959) examined the solution when q contains exactly two zeros, deriving full asymptotic series in terms of the Airy function through one transition point, and indicating the domain of validity of the various solutions obtained.

1.12 Other approximations

The W.K.B.J. solutions are the most common and the simplest approximations to which solutions of (1.1) are subject. On the other hand, improvements have been suggested by various authors. For example, Iami [61] (1948) has produced a better representation through the transition point, that reduces to Langer's if a certain parameter is large. Later, he [62] (1958) applied his improved technique to a problem relating to heat transfer in a lamina boundary layer, while Jenssen [69] (1960) used this improvement in discussing the asymptotic solution of the differential equation governing symmetrically loaded toroidal shells.

Bailey [5, 6] (1954) produced another improvement when considering the reflection of waves by an inhomogeneous medium. He replaced the W.K.B.J. solutions by an approximation valid through a transition point, but not of the Airy integral form. He calculated the reflection coefficients and also discussed an iterative process whereby the results could be further improved. Hines [59] (1953) also dealt with the reflection of waves from varying media by introducing a distinctive technique, while Moriguchi [82] (1960) considered the phase shift formula in the improved W.K.B.J. formula.

As far as the propagation of electromagnetic waves in an anisotropic ionosphere is concerned (see further, section 1.14), Försterling [34] (1942) produced equations of a 'coupled' type suitable for approximate solution by the method of variation of parameters. Gibbons and Nertney [42] (1951) produced a method for solving the left-hand side of these equations using a solution similar to (1.2) but with q replaced by p, where p is chosen to be such that (1.2) is now an exact solution to the equation. Smyth [101] (1952) objected to this procedure, observing that the properties of the medium were really modified so that the W.K.B.J. solutions became the exact solutions of the wave equation. This criticism is most unjust, and reveals lack of understanding of the method. The physical profile remained the same, but the solutions used were not W.K.B.J. solutions, although they reduced to the ordinary W.K.B.J. solutions in certain legitimate ranges. In spite of Nertney's [87] (1952) just reply, Smyth [102] (1952) persisted in a second letter to emphasize his point of view. Later, Gibbons and Nertney [43] (1952) applied the method of variation of parameters to include the coupling due to terms on the right-hand side of the equations, terms that were neglected in their (1951) paper.

1.13 Poles in the profile

In the previous sections, only zeros of q have been considered. Certain problems lead to poles in the functional form of q, and this case has received less attention. Gibbons and Schrag [44] (1952) have considered the problem of matching together the W.K.B.J. solutions on each side of the pole, while Heading [50] (1953) has given the Stokes constants for various cases. Budden's text [24] (pages 348–352, 474–481) may also be consulted.

1.14 Fourth-order equations

Anisotropic radio propagation is governed by fourth-order equations, and for high enough frequencies the W.K.B.J. technique may be applied in this case, although the mathematical justification of this process still appears to demand further investigation. Booker [14] (1938) used four solutions similar to primitive W.K.B.J. forms in his considerations of wave-packets and ray-paths, and Bremmer's contribution may be found in his text [19]. Eckersley [32] (1950) was the first to use his method of phase integration throughout the various Riemann sheets upon which was represented the more general function q involved. No proof of the method was offered, and the treatment was very unsatisfactory, undertaken to show that reflection would take place since at that time no analytical solution had been obtained by which the reflection processes could be examined. Rydbeck [96] (1951) employed the same technique to discuss the theory of 'magneto-ionic triple splitting', and Pfister [92] (1953) also dealt with the same subject. Budden [22] (1952) used the W.K.B.J. solutions in his theory of 'limiting polarization', a term that refers to the propagation of the characteristic waves towards the base of the ionosphere, and in (1955) he [23] employed the W.K.B.J. solutions deep into the ionosphere in order to provide the numerical basis for starting the numerical integration of the differential equations on an electronic computer.

1.15 Other types of equation

Phase-integral methods are standardized on account of the use made of the Airy equation; its solutions provide comparison functions by which the solutions of the original equation may be examined. Problems arise that lead to more complicated comparison equations, and provided that solutions can be found, often by contour-integral representations, progress can be made towards solving the original equation. Workers in certain branches of hydrodynamics have encountered various cases and we should mention the work of Heisenberg [58] (1924, before he became deeply immersed in his theory of matrix mechanics), Lin [78] (1945), Meksyn [80] (1946), Wasow [111, 112] (1950, 1953), Langer [77] (1957) and Rabenstein [93] (1958). Wasow has written [112] that 'complications more serious than those that occur in the asymptotic theory of certain second order differential equations of quantum mechanics' arise in this type of problem. Such considerations, it should be stated, do not form part of the object of our present text.

CHAPTER II

The W.K.B.J. Solutions

2.1 The Ricatti equation

The simplest derivation of the W.K.B.J. solutions originates from an appropriate change of the dependent variable. The given equation is

$$w'' + h^2 q(z, h) w = 0, \tag{2.1}$$

where the parameter h is taken to be large and positive, and where the function $q(z, h)$ tends to a limit as $h \to \infty$ for fixed complex z. A prime denotes differentiation with respect to z. If we let

$$w = \exp\left(\int^z \phi \, dz\right)$$

with a constant lower limit of integration, then

$$w' = \phi \exp\left(\int \phi \, dz\right)$$

and

$$w'' = \phi^2 \exp\left(\int \phi \, dz\right) + \phi' \exp\left(\int \phi \, dz\right),$$

yielding

$$\phi^2 + \phi' + h^2 q = 0.$$

If we now assume a development of ϕ in the form

$$\phi = \phi_0(z, h) h + \phi_1(z, h) + \phi_2(z, h) h^{-1} + \ldots,$$

where each $\phi_j(z, h)$ tends to a limit as $h \to \infty$ for fixed z, then substitution yields

$$\phi_0^2 h^2 + 2\phi_0 \phi_1 h + \phi_1^2 + 2\phi_0 \phi_2 + \ldots + \phi_0' h + \phi_1' + \ldots + h^2 q = 0 \tag{2.2}$$

as far as terms in h^2, h, h^0 are concerned. Equating coefficients (and this is not a unique process on account of the parameter h existing in all the ϕ_j), we obtain

$$\phi_0^2 + q = 0, \qquad 2\phi_0 \phi_1 + \phi_0' = 0, \qquad \phi_1^2 + 2\phi_0 \phi_2 + \phi_1' = 0.$$

Hence $\qquad \phi_0 = \pm iq^{1/2}, \qquad \phi_1 = -\phi_0'/2\phi_0.$

In terms of ϕ_0 and ϕ_1 the approximate solution for w is given by

$$
\left.\begin{aligned}
w &= \exp\left[\int (h\phi_0 + \phi_1)\,dz\right] \\
&= \exp\left(\int h\phi_0\,dz - \tfrac{1}{2}\log\phi_0\right) \\
&\propto q^{-1/4}\exp\left(\pm ih\int q^{1/2}\,dz\right).
\end{aligned}\right\} \qquad (2.3)
$$

The two solutions

$$
\begin{aligned}
w_1 &= q^{-1/4}\exp\left(ih\int q^{1/2}\,dz\right), \\
w_2 &= q^{-1/4}\exp\left(-ih\int q^{1/2}\,dz\right)
\end{aligned} \qquad (2.4)
$$

are known as the two W.K.B.J. approximations. The square root $q^{1/2}$ must be rendered definite by branch cuts in the complex z-plane radiating outwards from the zeros of q; the path of integration must not cross these cuts.

The derivative of w_1 appears to involve two terms, since a product is involved. But only that term arising from the differentiation of the exponential is retained, since the term obtained by differentiating the factor $q^{-1/4}$ would be modified by the ϕ_2 term if the development (2.3) were continued. Hence

$$
w_1' = ihq^{1/4}\exp\left(ih\int q^{1/2}\,dz\right), \qquad w_2' = -ihq^{1/4}\exp\left(-ih\int q^{1/2}\,dz\right)
$$

to this order of accuracy, the Wronskian being given by

$$
w_1 w_2' - w_1' w_2 = -2ih,
$$

provided the same lower limit is chosen for the integrals in both w_1 and w_2.

This simple treatment, though yielding the required approximate solutions, is not really adequate, since nothing is specified regarding the magnitude of the error involved in the approximation, nor about the domain of validity in the complex z-plane in which the solutions are valid. The solutions of equation (2.1) are single-valued through a domain free from singularities of $q(z, h)$, yet the approximate solutions (2.4) clearly cannot be single-valued owing to the roots of q involved. This demonstrates that the solutions (2.4) can only be valid in some restricted domain of the complex z-plane.

2.2 Comparison equations

A clearer idea of the approximations involved is obtained by comparing the given equation (2.1) with standard comparison equations possessing known solutions. The comparison equation is chosen to be

$$d^2 X/d\xi^2 = j(\xi) X,$$

where $j \equiv 1$ or ξ. The former is used if $q(z, h)$ does not vanish in the domain considered, while the latter is used if $q(z, h)$ vanishes, say at $z = 0$.

If $z = z(\xi)$, where ξ is a new independent variable,

$$dw/dz = w'/z',$$

where a prime now denotes differentiation with respect to ξ, and

$$\frac{d^2 w}{dz^2} = \frac{d}{d\xi}\left(\frac{w'}{z'}\right)\frac{1}{z'} = \frac{w''}{z'^2} - \frac{w' z''}{z'^3}.$$

The given equation (2.1) transforms to

$$w'' - (z''/z') w' + h^2 q z'^2 w = 0.$$

To eliminate the term in w', we transform the dependent variable thus: let

$$w = z'^{1/2} X,$$

yielding

$$w' = z'^{1/2} X' + \tfrac{1}{2} z'^{-1/2} z'' X,$$

$$w'' = z'^{1/2} X'' + z'^{-1/2} z'' X' + \tfrac{1}{2} z'^{-1/2} z''' X - \tfrac{1}{4} z'^{-3/2} z''^2 X.$$

Hence

$$X'' + h^2 q z'^2 X = \left(\frac{3}{4} \cdot \frac{z''^2}{z'^2} - \frac{1}{2} \cdot \frac{z'''}{z'}\right) X. \tag{2.5}$$

The left-hand side can now be made equivalent to the comparison equation by choosing $1 \equiv j = -h^2 q z'^2$ provided q does not vanish in the domain considered, yielding

$$\xi = \pm ih \int\limits_0^z q^{1/2} dz$$

where the lower limit, being arbitrary, is chosen to be that point at which q vanishes; the upper limit z must not of course vanish.

If the right-hand side of equation (2.5) is neglected, then

$$X'' = X,$$

so

$$X = \exp \xi$$

yielding

$$w = z'^{1/2} X \propto q^{-1/4} \exp\left(\pm ih \int\limits_0^z q^{1/2} dz \right),$$

namely the two W.K.B.J. solutions. By examining the coefficients of X in equation (2.5), we see that a *necessary* but not *sufficient* condition for this solution to be valid is that

$$|h^2 q z'^2| = 1 \gg \left| \frac{3}{4} \cdot \frac{z''^2}{z'^2} - \frac{1}{2} \cdot \frac{z'''}{z'} \right|. \tag{2.6}$$

This inequality may be expressed in terms of q and its derivatives with respect to z. If a prime attached to q denotes differentiation with respect to z, we have

$$\frac{dz}{d\xi} = \frac{i}{h} q^{-1/2},$$

$$\frac{d^2 z}{d\xi^2} = -\frac{i}{2h} q^{-3/2} q' \cdot \frac{i}{h} q^{-1/2} = \frac{1}{2h^2} q^{-2} q',$$

$$\frac{d^3 z}{d\xi^3} = \frac{1}{2h^2} (q^{-2} q'' - 2q^{-3} q'^2) \cdot \frac{i}{h} q^{-1/2},$$

so the right-hand bracket in equation (2.5) becomes

$$\frac{5}{16h^2} \cdot \frac{q'^2}{q^3} - \frac{1}{4h^2} \cdot \frac{q''}{q^2}. \tag{2.7}$$

In terms of q, inequality (2.6) takes the form

$$1 \gg \left| \frac{5}{16h^2} \cdot \frac{q'^2}{q^3} - \frac{1}{4h^2} \cdot \frac{q''}{q^2} \right|, \tag{2.8}$$

showing that h must be large and that q must not be too near zero. It will be seen later that the W.K.B.J. solutions are valid only in certain well-defined sectors around $z = 0$.

On the other hand, the left-hand side of equation (2.5) can be made equivalent to the comparison equation by choosing $\xi \equiv j = -h^2 q z'^2$, so

$$\xi^{1/2} d\xi = \pm i h q^{1/2} dz,$$

or

$$\xi = \left(\pm \frac{3ih}{2} \int_0^z q^{1/2} dz \right)^{2/3}, \tag{2.9}$$

where ξ vanishes when q vanishes. If the right-hand side of equation (2.5) is neglected, then

$$X'' = \xi X,$$

possessing the two standard solutions $\mathrm{Ai}\,(\xi)$ and $\mathrm{Bi}\,(\xi)$, being the Airy integral and its companion function which are studied in Chapter III and in the Appendix. This yields

$$w = z'^{1/2} X$$

$$\propto (\xi/q)^{1/4} X$$

$$\propto q^{-1/4} \left(\int_0^z q^{1/2} dz \right)^{1/6} \mathrm{Ai} \left[\left(\frac{3ih}{2} \int_0^z q^{1/2} dz \right)^{2/3} \right]. \tag{2.10}$$

It will be seen later that this approximate solution is not only valid in a domain completely surrounding $z = 0$, but also at $z = 0$ itself.

2.3 The integral equation

To demonstrate the nature of the approximations and of the errors involved, a simple integral equation must be formed for $X(\xi)$.

Let us write equation (2.5) in the form

$$X'' - jX = gX,$$

where $j \equiv 1$ or ξ, and where g stands for the bracket on the right-hand side. The integral equation, soluble by successive substitutions, is derived by the same technique as differential equations are solved by the method of variation of parameters.

If $Z_1(\xi)$ and $Z_2(\xi)$ are two solutions of the comparison equation $X'' - jX = 0$, let us seek to express $X(\xi)$ in the form

$$X(\xi) = \alpha(\xi) Z_1(\xi) + \beta(\xi) Z_2(\xi).$$

Since two unknown functions $\alpha(\xi)$ and $\beta(\xi)$ are involved, and only one equation must be satisfied, any useful relationship may be postulated between $\alpha(\xi)$ and $\beta(\xi)$. It follows upon differentiation that

$$X' = \alpha' Z_1 + \alpha Z_1' + \beta' Z_2 + \beta Z_2'.$$

Without loss of generality, we now choose for convenience

$$\alpha' Z_1 + \beta' Z_2 = 0; \tag{2.11}$$

then
$$\begin{aligned} X'' &= \alpha' Z_1' + \alpha Z_1'' + \beta' Z_2' + \beta Z_2'' \\ &= j(\alpha Z_1 + \beta Z_2) + gX. \end{aligned}$$

Since $Z_1'' - jZ_1 = 0 = Z_2'' - jZ_2$, this reduces to

$$\alpha' Z_1' + \beta' Z_2' = gX.$$

This equation, together with (2.11), yields

$$\alpha' = gZ_2 X/(Z_1' Z_2 - Z_1 Z_2'),$$

$$\beta' = -gZ_1 X/(Z_1' Z_2 - Z_1 Z_2'),$$

or
$$\alpha(\xi) = \frac{1}{W} \int^{\xi} g(t) Z_2(t) X(t) \, dt + A,$$

$$\beta(\xi) = -\frac{1}{W} \int^{\xi} g(t) Z_1(t) X(t) \, dt + B,$$

where the Wronskian $W \equiv Z_1' Z_2 - Z_1 Z_2'$ is a constant for comparison equations not containing an X' term. A and B are arbitrary constants. Hence

$$X(\xi) = AZ_1(\xi) + BZ_2(\xi) + \frac{1}{W} \int^{\xi} [Z_1(\xi)Z_2(t) - Z_2(\xi)Z_1(t)]g(t)X(t)\,dt.$$
(2.12)

This is the required integral equation, since X occurs both outside and inside the sign of integration. A change of the lower limit of integration merely changes the values of the arbitrary constants A and B.

2.4 Approximate solutions of the intregal equation

If $j \equiv 1$, we have $Z_1(\xi) = e^{\xi}$, $Z_2(\xi) = e^{-\xi}$, $W = 2$, so equation (2.12) becomes

$$X(\xi) = Ae^{\xi} + Be^{-\xi} + \tfrac{1}{2} \int_C^{\xi} (e^{\xi - t} - e^{-\xi + t})g(t)X(t)\,dt, \quad (2.13)$$

where $t = C$ is some point in the complex t-plane.

We shall first of all solve this equation by successive approximations with $A = 1$, $B = 0$. The first approximation to $X(\xi)$ is e^{ξ}, ignoring the effect of the integral. Such a solution is exponentially large if $\mathrm{R1}\,\xi > 0$ but exponentially small if $\mathrm{R1}\,\xi < 0$. In the former case, e^{ξ} is said to be *dominant*, while in the latter case it is *subdominant*; we shall consider the former case when e^{ξ} is dominant.

The next approximation is obtained by substituting $X(t) = e^t$ into the integrand of (2.13), yielding

$$X(\xi) = e^{\xi} + \tfrac{1}{2} \int_C^{\xi} (e^{\xi} - e^{-\xi + 2t})g(t)\,dt,$$

provided the magnitude of this correction term is much smaller than that of the first approximation $|e^{\xi}|$. Its magnitude is less than

$$\max \tfrac{1}{2}|e^{\xi} - e^{-\xi + 2t}| \cdot \int_C^{\xi} |g(t)\,dt|.$$

Now from (2.7), $|g(t)\,dt|$ in the ξ-plane equals, in the z-plane,

$$\frac{1}{h^2}\left|\left(\frac{5}{16} \cdot \frac{q'^2}{q^3} - \frac{1}{4} \cdot \frac{q''}{q^2}\right)q^{1/2}h\,dz\right|.$$

Provided that C and ξ are not too near any point at which q vanishes, the integral of the modulus of this expression between any point C and any point ξ is usually less than M/h, where M is a constant independent of C, ξ and h. If q contains h as previously described in section 2.1, this result may only be true for $h > h_0$.

If this is so, the error is less in magnitude than

$$(M/h) \max \tfrac{1}{2} |e^{\xi} - e^{-\xi + 2t}|.$$

We now restrict the domain in which C and ξ are allowed to be situated by asserting that $\mathrm{Rl}\, t$ must increase along the path from C to ξ. Then

$$
\begin{aligned}
|e^{\xi} - e^{-\xi + 2t}| &\leqslant |e^{\xi}| + |e^{-\xi + 2t}| \\
&= e^{\mathrm{Rl}\,\xi} + e^{-\mathrm{Rl}\,\xi} e^{2\,\mathrm{Rl}\,t} \\
&\leqslant e^{\mathrm{Rl}\,\xi} + e^{-\mathrm{Rl}\,\xi} e^{2\,\mathrm{Rl}\,\xi} \\
&= 2e^{\mathrm{Rl}\,\xi}.
\end{aligned}
$$

Hence the first correction term is less than $|Me^{\xi}/h|$, so the development of the solution appears as

$$X(\xi) = e^{\xi}\left[1 + O\!\left(\frac{1}{h}\right)\right]$$

valid at all points to the right of C, such that $\mathrm{Rl}\,\xi \geqslant \mathrm{Rl}\,C$.

The largest domain to the right of C is obtained if we choose C (which has been arbitrary until now) to be near $\xi = 0$ on its right, as shown in Fig. 1. The boundaries of this domain in the ξ-plane are given by $\mathrm{Rl}\,\xi = 0$; in the z-plane, these lines are defined by

$$\mathrm{Im} \int_0^z q^{1/2}\, dz \tag{2.14}$$

and are termed *anti-Stokes lines*. Several such lines radiate outwards from the point $z = 0$, which is usually called a *transition point*. If q is approximately linear near $z = 0$, say $q \doteqdot z$ (though a complex constant factor may also be inserted if required), then these anti-Stokes lines in the immediate neighbourhood of $z = 0$ are given by

$$\text{Im} \int_0^z z^{1/2} \, dz = \text{Im} \tfrac{2}{3} z^{3/2} = 0.$$

If $z = re^{i\theta}$, the lines locally are given by $\sin\tfrac{3}{2}\theta = 0$, so $\theta = 0$, $\pm \tfrac{2}{3}\pi$, yielding three distinct lines forming three sectors.

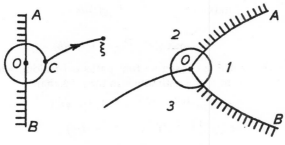

Fig. 1. ξ-plane z-plane

Hence the approximate solution of (2.13) is $X = e^{\xi}$, uniformly valid through domain 1 only, that is, just in that domain for which it is dominant. In terms of z,

$$w = q^{-1/4} \exp\left(ih \int_0^z q^{1/2} \, dz \right)$$

where the branch of $q^{1/2}$ is such that the exponential is dominant in domain 1. This approximation is *uniformly asymptotic*, in the sense that the error is $O(1/h)$ times the main approximation, and that this is maintained throughout the domain. It should be stressed that the approximation is *not* valid on the boundary of the domain, and that the above discussion has *not* dealt with the situation further away from the transition point when other zeros of q must be taken into account.

Similarly, if $A = 0$, $B = 1$ in equation (2.13), consider a solution

$X(\xi) = e^{-\xi}$ subdominant for $\mathrm{Rl}\,\xi > 0$ in domain 1. The first correction term yields

$$X(\xi) = e^{-\xi} + \tfrac{1}{2} \int\limits_{C}^{\xi} (e^{\xi - 2t} - e^{-\xi}) g(t)\,dt,$$

provided this correction term is much smaller in magnitude than the first approximation $|e^{-\xi}|$. Its magnitude is less than

$$\max \tfrac{1}{2} |e^{\xi - 2t} - e^{-\xi}| \int\limits_{C}^{\xi} |g(t)\,dt|$$

$$\leqslant \max \tfrac{1}{2} |e^{\xi - 2t} - e^{-\xi}| \,(M/h)$$

as before. Now if ξ is chosen so that paths exist along which $\mathrm{Rl}\,t$ decreases from C to ξ, this estimate is in turn less than

$$\max \tfrac{1}{2} (e^{\mathrm{Rl}\,\xi} e^{-2\,\mathrm{Rl}\,t} + e^{-\mathrm{Rl}\,\xi})(M/h)$$

$$\leqslant \tfrac{1}{2} (e^{\mathrm{Rl}\,\xi} e^{-2\,\mathrm{Rl}\,\xi} + e^{-\mathrm{Rl}\,\xi})(M/h)$$

$$= e^{-\mathrm{Rl}\,\xi}(M/h)$$

$$= |e^{-\xi}|\, M/h.$$

Hence, under these circumstances, the development of the solution appears as

$$X(\xi) = e^{-\xi} \left[1 + O\!\left(\frac{1}{h}\right) \right].$$

If C is chosen so that $\mathrm{Rl}\,C$ is as large as possible, paths may be drawn from C to all points ξ on its left as in Fig. 2, such that $\mathrm{Rl}\,t$ decreases along the path. Paths may cross the lines OA, OB, to all points in the domain such that $-\tfrac{3}{2}\pi < \arg \xi < \tfrac{3}{2}\pi$, the boundaries being OA' attained from OB by a rotation of π clockwise, and OB' attained from OA by a rotation of π anticlockwise. Two sheets are necessary in the complex ξ-plane to exhibit this.

In the z-plane, these lines are given by the anti-Stokes lines

$$\mathrm{Rl}\,\xi = \mathrm{Im} \int\limits_{0}^{z} q^{1/2}\,dz = 0,$$

three of which emerge from O, with OA' and OB' coinciding. The branch cut is usually placed along this line. The solution $e^{-\xi}$ is uniformly valid all round $z = 0$ except on the line $A'OB'$ and of course near O itself.

This theory, following and extending that given by Jeffreys [66], does not demonstrate how far outwards these uniformly asymptotic solutions may be extended. Further zeros of q complicate the solution,

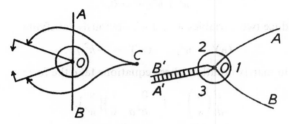

FIG. 2. ξ-plane z-plane

but a detailed examination of this problem has been given by Heading [56]. The solution just obtained

$$w(z) = q^{-1/4} \exp\left(- ih \int\limits_0^z q^{1/2}\, dz\right)$$

is valid in the domain AOB in which it is subdominant; it may be extended over the anti-Stokes lines OA and OB up to but not including the neighbouring anti-Stokes lines.

From now on, we shall omit the correction terms $e^{\pm\xi} O(1/h)$, their presence being implicitly assumed. Moreover, we shall only be concerned with functions q for which

$$\int\limits_C^\xi |g(t)\, dt| < M/h$$

is satisfied without further examination, for all points C and ξ in the domain under consideration.

If $j \equiv \xi$, the analysis is more complicated; details have been given by Jeffreys [66] and more fully by Heading [56]. Solution (2.10) now turns out to be uniformly asymptotic all round $z = 0$ and also at $z = 0$ itself.

2.5 The matrix method

One second-order linear differential equation may be expressed in terms of two first-order simultaneous differential equations. If

$$w'' + h^2 q w = 0,$$

we introduce two variables w and w', obviously satisfying

$$(w)' = w', \qquad (w')' = -h^2 q w.$$

Written in matrix notation, these equations take the form

$$\frac{d}{dz}\begin{pmatrix} w \\ w' \end{pmatrix} = \begin{pmatrix} 0 & 1 \\ -h^2 q & 0 \end{pmatrix}\begin{pmatrix} w \\ w' \end{pmatrix}. \tag{2.15}$$

If the column matrix is denoted by $\mathbf{w}$ and the 2×2 matrix by $\mathbf{T}$, we have

$$\mathbf{w}' = \mathbf{T}\mathbf{w}.$$

The two variables w and w' are linked by the simultaneous character of the equations, in such a way that no approximate solution is obvious.

But the simultaneous character of the equations may be eliminated *as far as possible* by changing the dependent column $\mathbf{w}$ to a new column $\mathbf{f}$. Let

$$\mathbf{w} = \mathbf{R}\mathbf{f},$$

namely

$$\begin{pmatrix} w \\ w' \end{pmatrix} = \begin{pmatrix} r_{11} & r_{12} \\ r_{21} & r_{22} \end{pmatrix}\begin{pmatrix} f_1 \\ f_2 \end{pmatrix}, \tag{2.16}$$

where $\mathbf{R}$, with variable elements, remains to be chosen.

It follows that

$$\mathbf{w}' = \mathbf{R}\mathbf{f}' + \mathbf{R}'\mathbf{f},$$

and substitution yields

$$\mathbf{R}\mathbf{f}' = \mathbf{T}\mathbf{R}\mathbf{f} - \mathbf{R}'\mathbf{f},$$

or

$$\mathbf{f}' = \mathbf{R}^{-1}\mathbf{T}\mathbf{R}\mathbf{f} - \mathbf{R}^{-1}\mathbf{R}'\mathbf{f},$$

provided $\mathbf{R}$ is non-singular, that is, if $\det \mathbf{R} \neq 0$. The matrix $\mathbf{R}$ is now chosen so that $\mathbf{R}^{-1}\mathbf{T}\mathbf{R}$ is a diagonal matrix of order 2.

The characteristic (or latent) roots λ_1, λ_2 of $\mathbf{T}$ are the roots of the equation

$$\det(\mathbf{T} - \lambda\mathbf{I}) = 0,$$

namely of

$$\begin{vmatrix} -\lambda & 1 \\ -h^2 q & -\lambda \end{vmatrix} = 0,$$

reducing to

$$\lambda^2 = -h^2 q.$$

To each root $ihq^{1/2}$, $-ihq^{1/2}$ corresponds a characteristic vector $\mathbf{v}_1$, $\mathbf{v}_2$ respectively, namely

$$\mathbf{v}_1 = \begin{pmatrix} 1 \\ ihq^{1/2} \end{pmatrix}, \qquad \mathbf{v}_2 = \begin{pmatrix} 1 \\ -ihq^{1/2} \end{pmatrix}.$$

Then standard matrix theory proves that the matrix

$$\mathbf{R} = (\mathbf{v}_1 \mathbf{v}_2) = \begin{pmatrix} 1 & 1 \\ ihq^{1/2} & -ihq^{1/2} \end{pmatrix}$$

is such that $\quad \mathbf{R}^{-1}\mathbf{T}\mathbf{R} = \begin{pmatrix} ihq^{1/2} & 0 \\ 0 & -ihq^{1/2} \end{pmatrix} = \mathbf{D},$

say, a diagonal matrix. $\mathbf{R}$ is evidently non-singular if the two roots are distinct, namely if $q \neq 0$. Quite apart from the formal theory that applies equally to $n \times n$ matrices, the reader may easily check by direct multiplication of the 2×2 matrices that

$$\mathbf{R}^{-1}\mathbf{T}\mathbf{R} = \frac{\begin{pmatrix} -ihq^{1/2} & -1 \\ -ihq^{1/2} & 1 \end{pmatrix}}{-2ihq^{1/2}} \begin{pmatrix} 0 & 1 \\ -h^2 q & 0 \end{pmatrix} \begin{pmatrix} 1 & 1 \\ ihq^{1/2} & -ihq^{1/2} \end{pmatrix}$$

$$= \begin{pmatrix} ihq^{1/2} & 0 \\ 0 & -ihq^{1/2} \end{pmatrix}.$$

The transformed equation now takes the form

$$\begin{pmatrix} f_1 \\ f_2 \end{pmatrix}' = \begin{pmatrix} ihq^{1/2} & 0 \\ 0 & -ihq^{1/2} \end{pmatrix} \begin{pmatrix} f_1 \\ f_2 \end{pmatrix} - \frac{1}{-2ihq^{1/2}} \begin{pmatrix} -ihq^{1/2} & -1 \\ -ihq^{1/2} & 1 \end{pmatrix} \times$$

$$\times \begin{pmatrix} 0 & 0 \\ \frac{1}{2}ihq^{-1/2}q' & -\frac{1}{2}ihq^{-1/2}q' \end{pmatrix} \begin{pmatrix} f_1 \\ f_2 \end{pmatrix}$$

$$= \begin{pmatrix} ihq^{1/2} & 0 \\ 0 & -ihq^{1/2} \end{pmatrix} \begin{pmatrix} f_1 \\ f_2 \end{pmatrix} - \frac{q'}{4q} \begin{pmatrix} 1 & -1 \\ -1 & 1 \end{pmatrix} \begin{pmatrix} f_1 \\ f_2 \end{pmatrix}. \tag{2.17}$$

This matrix equation still represents two simultaneous equations for f_1 and f_2, but it now exhibits a very distinctive form. If we consider values of z at which $|q|$ is not small, then the terms in the last matrix product are small in magnitude compared with the terms containing the explicit large parameter h originating from the diagonal matrix. The terms involving f_2 in the first equation and f_1 in the second equation are called *coupling terms*, while the two equations themselves are known as *coupled equations*. The usual argument for solving these equations is as follows (see, for example, Budden [24], page 402):

If it is legitimate to neglect the coupling terms, the matrix equation reduces to two independent first order quations in f_1 and f_2, namely

$$f_1' = (ihq^{1/2} - q'/4q)f_1,$$

$$f_2' = (-ihq^{1/2} - q'/4q)f_2,$$

with the solutions

$$f_1 = Aq^{-1/4}\exp\left(ih\int q^{1/2}\,dz\right), \tag{2.18}$$

$$f_2 = Bq^{-1/4}\exp\left(-ih\int q^{1/2}\,dz\right). \tag{2.19}$$

These may be termed the two independent *characteristic waves* propagated in the medium, and appear to be valid approximations provided the coupling between them is neglected. It is obvious that coupling is not small at points where $q = 0$; the independent existence of f_1 and f_2 is no longer maintained at such points. These are called *transition points* in mathematical language and *reflection points* (or, more generally, *coupling points*) in physical language.

Equation (2.16) finally shows that

$$w = f_1 + f_2,$$

so this matrix method has produced the W.K.B.J. solutions simply and directly.

On the other hand, it should be pointed out that the simplicity of the equations (2.17) is deceptive. These equations should be solved by the same technique as used in section 2.4. It would then be found that if $\exp(ih \int q^{1/2} dz)$ is dominant, then f_1 has the form (2.18) while $f_2 = 0$ to this approximation. The error term would be $q^{-1/4} \exp(ih \int q^{1/2} dz) O(1/h)$ in both f_1 and f_2, the subdominant form (2.19) not entering the expression for f_2 at all.

Similarly, a second solution, subdominant this time, would be given by $f_1 = 0$, $f_2 = q^{-1/4} \exp(-ih \int q^{1/2} dz)$, the error term being $O(1/h)$ times this subdominant solution in both f_1 and f_2. Only if z is chosen to lie on lines such as $\operatorname{Im} \int q^{1/2} dz = 0$ can both f_1 and f_2 exist together in the respective forms given by (2.18) and (2.19).

2.6 Jeffrey's form

Jeffreys [66] has usually considered an equation of the form

$$w'' + [h^2 q_0(z) + h q_1(z) + q_2(z)] w = 0,$$

though this form does not always correspond to those occurring in physical problems. Equation (2.5) may then be written as

$$X'' + (h^2 q_0 + h q_1) z'^2 X = \left(-q_2 z'^2 + \frac{3}{4} \cdot \frac{z''^2}{z'^2} - \frac{1}{4} \cdot \frac{z'''}{z'} \right) X,$$

or even as

$$X'' + \left(h^2 q_0 + h q_1 + \frac{q_1^2}{4 q_0} \right) z'^2 X = \left[\left(\frac{q_1^2}{4 q_0} - q_2 \right) z'^2 + \frac{3}{4} \cdot \frac{z''^2}{z'^2} - \frac{1}{4} \cdot \frac{z'''}{z'} \right] X.$$

Jeffreys has usually recommended the relation

$$(h^2 q_0 + h q_1 + q_1^2/4 q_0) z'^2 = -1,$$

yielding

$$d\xi = ih q_0^{1/2} (1 + q_1/2 h q_0) dz,$$

so

$$\xi = ih \int^z q_0^{1/2} (1 + q_1/2 h q_0) dz.$$

The approximate solutions $X = e^{\pm \xi}$, uniformly asymptotic in the same sense as explained in section 2.4, are preferable to those solutions derived in that section under certain exceptional circumstances. The disadvantage is that the integrand now involved in ξ is not expressed directly in terms of q, but only in a somewhat mutilated form of q. Physically, q is related to the refractive index of a medium, and whenever possible it is preferable to use the refractive index itself rather than an approximation to it.

2.7 Notation and examples

To facilitate the continual writing and printing of expressions such as (2.18) and (2.19) complete with the end points of the paths of integration, we adopt the following simple yet useful notation.

By means of branch cuts, $\arg q$ is specified uniquely throughout the domain of interest in the z-plane; $q^{1/2}$ and $q^{1/4}$ are therefore uniquely specified. We use the symbol (a, z) to denote the W.K.B.J. solution (2.18), namely

$$(a, z) \equiv q^{-1/4} \exp\left(ih \int_a^z q^{1/2} dz\right), \tag{2.20}$$

where the path of integration does not cross the branch cuts. The order of the limits of integration is indicated by the order of the symbols in the bracket (a, z). Similarly,

$$(z, a) \equiv q^{-1/4} \exp\left(-ih \int_a^z q^{1/2} dz\right), \tag{2.21}$$

since the minus sign merely reverses the path of integration.

If z is such that the exponential in (2.20) is dominant, a suffix d is inserted, namely $(a, z)_d$; solution (2.21) is then subdominant, and a suffix s is used, namely $(z, a)_s$. If $\text{Im} \int_a^z q^{1/2} \, dz = 0$, the exponential is neither dominant nor subdominant; its magnitude may be termed *neutral*, and no suffix is used to denote this important physical property.

Sometimes, both limits of integration are constants. We use square brackets to denote the exponential of such an integral *with no $q^{-1/4}$ factor implied*, namely

$$[a, b] \equiv \exp\left(ih \int_a^b q^{1/2} dz\right).$$

Note that

$$[a, b][b, a] = 1,$$

$$(a, z)(z, a) = q^{-1/2}.$$

If the lower limit a is changed to b in $(a, z)_d$, the identity

$$q^{-1/4} \exp\left(ih \int_a^z q^{1/2} dz\right) = q^{-1/4} \exp\left(ih \int_b^z q^{1/2} dz + ih \int_a^b q^{1/2} dz\right)$$

is obviously written as

$$(a, z)_d = [a, b](b, z)_d,$$

showing the ease by which such expressions may be manipulated. Similarly,

$$(z, a)_s = (z, b)_s [b, a].$$

Sometimes, the change of limit may change dominancy into subdominancy and *vice versa*.

Example 1. If $z = x + iy$, and

$$w'' + h^2(1 - e^{kz}) w = 0, \qquad (k \text{ real and positive})$$

then $q \equiv 1 - e^{kz}$ is real all along the x-axis; we note that transition points occur when $1 - e^{kz}$ vanishes, namely when

$$kz = 2\pi ni \qquad (n \text{ integral})$$

or

$$z = 2\pi ni/k.$$

On the real axis, only one transition point occurs when $z = 0$. Let a branch cut be inserted between, say, $\arg z = 0$ and $\arg z = -\pi/6$, and let $\arg q = \pi$ when $x > 0$ and $\arg q = 2\pi$ when $x < 0$ along the real axis. Then

$$(0, x)_s = [e^{i\pi}(e^{kx} - 1)]^{-1/4} \exp\left(ih \int_0^x \sqrt{[e^{i\pi}(e^{kx} - 1)]}\, dx\right), \qquad x > 0$$

$$(0, x) = [e^{2i\pi}(1 - e^{kx})]^{-1/4} \exp\left(ih \int_0^x \sqrt{[e^{2i\pi}(1 - e^{kx})]}\, dx\right). \qquad x < 0$$

If $\sqrt{(e^{kx} - 1)} = v$, the first integral becomes, for $x > 0$,

$$-h \int_0^x \sqrt{(e^{kx} - 1)}\, dx = -\frac{2h}{k} \int \frac{v^2\, dv}{v^2 + 1} = -\frac{2h}{k}[v + \tan^{-1} v]$$

$$= -\frac{2h}{k}\left[\sqrt{(e^{kx} - 1)} + \tan^{-1}\sqrt{(e^{kx} - 1)}\right]_0^x$$

$$= -(2h/k)[\sqrt{(e^{kx} - 1)} + \tan^{-1}\sqrt{(e^{kx} - 1)}]$$

$$\doteq -(2h/k)(e^{kx/2} + \tfrac{1}{2}\pi)$$

approximately when $x \geqslant 0$, but there is of course no necessity for x to be large in the W.K.B.J. solutions.

Similarly, for $x < 0$ and using $v = \sqrt{(1 - e^{kx})}$, we obtain

$$-ih \int_0^x \sqrt{(1 - e^{kx})}\, dx = (2ih/k)\{-\sqrt{(1 - e^{kx})} -$$

$$-\tfrac{1}{2}kx + \log[1 + \sqrt{(1 - e^{kx})}]\}.$$

When x is large and negative, e^{kx} is small, so the integral has the approximate value

$$(2ih/k)(-1 + \log 2) - ihx.$$

Hence, when $|x|$ is large,

$$(0, x)_s = e^{-i\pi/4} e^{-kx/4} \exp[-h(2e^{kx/2} + \pi)/k], \qquad x > 0$$

$$(0, x) = -i \exp[2ih(-1 + \log 2)/k] e^{-ihx}. \qquad x < 0$$

The first solution represents an *evanescent* wave in an overdense medium, while the second solution represents a *progressive* wave in a homogeneous medium for which $q = 1$.

It should be pointed out that the two forms do not represent one solution all along the x-axis.

In Fig. 3, anti-Stokes lines are represented by full lines, and the branch cut by the wavy line. The subdominant solution in domain 1 is also valid in 2 and 3 but not on the line OA, while the solution neutral on OA and subdominant in 3 is valid also in 1 and 2 but not on the line OB, in keeping with the rules established in section 2.4. The effect of the branch cut will be discussed in Chapter III.

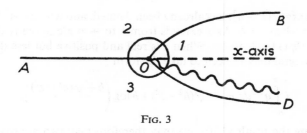

Fig. 3

Example 2. We shall now obtain approximate expressions for the Hankel functions. If we substitute $u = wz^{-1/2}$ in Bessel's equation of order h (a symbol used to be consistent with our previous notation)

$$u'' + z^{-1}u' + (1 - h^2 z^{-2})u = 0,$$

we obtain

$$w'' + (1 - H^2 z^{-2})w = 0,$$

where $H^2 = h^2 - \frac{1}{4}$. This is similar to equation (2.1), with q equal to $H^{-2} - z^{-2}$. Since $H = h[1 + O(1/h^2)]$, we shall replace H by h since h is assumed real, large and positive. We shall take $\arg(1 - h^2/z^2)$ to be zero along the positive x-axis from the transition point $z = h$.

To avoid unnecessary constant factors, we shall now write

$$q = 1 - h^2/z^2,$$

implying that

$$q = e^{i\pi}(h^2/z^2 - 1)$$

on the positive real axis between 0 and h, if a branch cut is inserted from $-i\infty$ to h. To evaluate $\int q^{1/2} dz$, we substitute $z = h \sec \theta$, yielding

$$i \int_{h}^{z} q^{1/2} dz = i \int \sin \theta \cdot h \sec \theta \tan \theta \cdot d\theta$$

$$= ih(\tan \theta - \theta)$$

$$= i\left[\sqrt{(z^2 - h^2)} - h \cos^{-1}(h/z) \right]_{h}^{z}$$

$$= i[\sqrt{(z^2 - h^2)} - h \cos^{-1}(h/z)],$$

where $\arg \sqrt{(z^2 - h^2)}$ has already been defined, and where $\cos^{-1}(h/z)$ is zero when $z = h$. As x extends from h to $+\infty$ along the real axis, $\cos^{-1}(h/x)$ tends to $\frac{1}{2}\pi$. When z is real and positive but less than h, this integral may be seen to take the form

$$i \int_{h}^{z} q^{1/2} dz = -\sqrt{(h^2 - z^2)} + h \log \left(\frac{h + \sqrt{(h^2 - z^2)}}{z} \right).$$

Using the result (2.10), we may therefore take two approximate solutions of Bessel's equation to be

$$\left. \begin{matrix} u_1 \\ u_2 \end{matrix} \right\} = wz^{-1/2} = z^{-1/2}(1 - h^2/z^2)^{-1/4} [\sqrt{(z^2 - h^2)} - h \cos^{-1}(h/z)]^{1/6} \times$$

$$\times \text{Ai} \left\{ \left[\pm \tfrac{3}{2} i (\sqrt{(z^2 - h^2)} - h \cos^{-1} \tfrac{h}{z}) \right]^{2/3} \right\}, \qquad (2.22)$$

where u_1 and u_2 refer to the plus and minus sign respectively. This solution is valid all along the positive x-axis, excluding $x = 0$ but including the transition point $x = h$. It is also valid in a domain bordering the real x-axis, but the theory as developed in this chapter does not enable us to state how far outwards this domain of validity extends.

The general solution of Bessel's equation consists of a linear combination of these two solutions (2.22). We seek the appropriate linear combinations so that the solution represents approximately the Hankel functions $H_h^{(1)}(z)$ and $H_h^{(2)}(z)$. To compare these functions with the approximate solutions (2.22), we compare their asymptotic expressions along the real z-axis for $x \gg h$. Along this axis, the exponentials occurring in these expressions both for the Hankel functions and for the Airy integral are neutral in magnitude, so there is no danger of confusion that would arise if one solution were dominant.

We shall carry out the identification when $\mathrm{Rl}\, z = x \gg h$. We recall that

$$H_h^{(1)}(x) \sim \sqrt{(2/\pi)}\, x^{-1/2} \exp\left[i(x - \tfrac{1}{2}h\pi - \tfrac{1}{4}\pi)\right],$$
$$H_h^{(2)}(x) \sim \sqrt{(2/\pi)}\, x^{-1/2} \exp\left[-i(x - \tfrac{1}{2}h\pi - \tfrac{1}{4}\pi)\right]$$

(see Watson [114], Chapter VII), and that

$$\mathrm{Ai}\,(\zeta) \sim \tfrac{1}{2}\pi^{-1/2}\,\zeta^{-1/4}\exp\left(-\tfrac{2}{3}\zeta^{3/2}\right), \tag{2.23}$$

provided $-\pi < \arg \zeta < \pi$ (see Chapter III and the Appendix).

When $x \gg h$, we have

$$\sqrt{(x^2 - h^2)} - h\cos^{-1}(h/x) \doteq x - \tfrac{1}{2}h\pi,$$

and taking $\pm i = e^{\pm i\pi/2}$ in (2.22) in order to be definite,

$$\arg\left[\tfrac{3}{2}e^{\pm i\pi/2}\left(\sqrt{(x^2 - h^2)} - h\cos^{-1}\frac{h}{x}\right)\right]^{2/3} = \pm \tfrac{1}{3}\pi,$$

both values lying within the range of validity of expansion (2.23). Hence

$$\left.\begin{array}{c}u_1\\u_2\end{array}\right\} \sim x^{-1/2}(x - \tfrac{1}{2}h\pi)^{1/6} \cdot \tfrac{1}{2}\pi^{-1/2} \times$$
$$\times \left[\tfrac{3}{2}e^{\pm i\pi/2}(x - \tfrac{1}{2}h\pi)\right]^{-1/6} \exp\left[-e^{\pm i\pi/2}(x - \tfrac{1}{2}h\pi)\right]$$
$$= \tfrac{1}{2}\pi^{-1/2}(\tfrac{3}{2}e^{\pm i\pi/2})^{-1/6} x^{-1/2} \exp\left[-e^{\pm i\pi/2}(x - \tfrac{1}{2}h\pi)\right].$$

By comparing the coefficients of $e^{\pm ix}$ in these expressions with those occurring in the asymptotic expressions of the Hankel functions, we see that $H_h^{(1)}(x)$ is expressed solely in terms of u_2 and $H_h^{(2)}(x)$ in terms of u_1. If

$$H_h^{(1)}(x) = A_2 u_2, \qquad H_h^{(2)}(x) = A_1 u_1,$$

we have, for $H_h^{(1)}(x)$,

$$\sqrt{(2/\pi)} \exp\left[i\left(-\tfrac{1}{2}h\pi - \tfrac{1}{4}\pi\right)\right] =$$
$$A_2 \cdot \tfrac{1}{2}\pi^{-1/2}\left(\tfrac{3}{2}e^{-i\pi/2}\right)^{-1/6} \exp\left(-\tfrac{1}{2}ih\pi\right),$$

and for $H_h^{(2)}(x)$

$$\sqrt{(2/\pi)} \exp\left[-i\left(-\tfrac{1}{2}h\pi - \tfrac{1}{4}\pi\right)\right] =$$
$$A_1 \cdot \tfrac{1}{2}\pi^{-1/2}\left(\tfrac{3}{2}e^{i\pi/2}\right)^{-1/6} \exp\left(\tfrac{1}{2}ih\pi\right),$$

yielding

$$A_2 = 2^{4/3}\, 3^{1/6}\, e^{-i\pi/3}, \qquad A_1 = 2^{4/3}\, 3^{1/6}\, e^{i\pi/3}.$$

Hence, all along the positive x-axis but excluding the neighbour-hood of the singularity $x = 0$, we have the approximate solutions

$$H_h^{(1)}(x) = 2^{4/3}\, 3^{1/6}\, e^{-i\pi/3}(x^2 - h^2)^{-1/4} \times$$
$$\times \left[\sqrt{(x^2 - h^2)} - h\cos^{-1}(h/x)\right]^{1/6} \times$$
$$\times \mathrm{Ai}\left\{\left[\tfrac{3}{2}e^{-i\pi/2}\left(\sqrt{(x^2 - h^2)} - h\cos^{-1}\frac{h}{x}\right)\right]^{2/3}\right\},$$

$$H_h^{(2)}(x) = 2^{4/3}\, 3^{1/6}\, e^{i\pi/3}(x^2 - h^2)^{-1/4} \times$$
$$\times \left[\sqrt{(x^2 - h^2)} - h\cos^{-1}(h/x)\right]^{1/6} \times$$
$$\times \mathrm{Ai}\left\{\left[\tfrac{3}{2}e^{i\pi/2}\left(\sqrt{(x^2 - h^2)} - h\cos^{-1}\frac{h}{x}\right)^{2/3}\right]\right\}.$$

Such approximations, though perhaps in different forms, have often been used; see, for example, Bremmer [19], *Terrestrial Radio Waves*.

In particular, we may consider the values of these functions at the transition point $x = h$. In evaluating such expressions, we note that

$$(x^2 - h^2)^{-1/4}\left[\sqrt{(x^2 - h^2)} - h\cos^{-1}(h/x)\right]^{1/6}$$
$$= \left\{\epsilon^{-3}\left[\epsilon - h\tan^{-1}(\epsilon/h)\right]\right\}^{1/6} \qquad \left[\text{where } \epsilon = \sqrt{(x^2 - h^2)}\right]$$
$$= \left\{\epsilon^{-3}\left[\epsilon - h(\epsilon/h - \epsilon^3/3h^3 + \ldots)\right]\right\}^{1/6}$$
$$\to (1/3h^2)^{1/6}$$

as $\epsilon \to 0$. It follows that

$$H_h^{(1)}(h) = 2^{4/3} 3^{1/6} e^{-i\pi/3} (3h^2)^{-1/6} \operatorname{Ai}(0),$$

$$H_h^{(2)}(h) = 2^{4/3} 3^{1/6} e^{i\pi/3} (3h^2)^{-1/6} \operatorname{Ai}(0),$$

where $\operatorname{Ai}(0) = 3^{-2/3}/\Gamma(\tfrac{2}{3})$ (see Miller [81] and the Appendix). The Bessel function of order h is then found as follows:

$$\begin{aligned}
J_h(h) &= \tfrac{1}{2}[H_h^{(1)}(h) + H_h^{(2)}(h)] \\
&= 2^{4/3} 3^{1/6} \cos \tfrac{1}{3}\pi \, (3h^2)^{-1/6} \operatorname{Ai}(0) \\
&= 2^{1/3} 3^{-2/3} h^{-1/3}/\Gamma(\tfrac{2}{3}).
\end{aligned}$$

But $\Gamma(\tfrac{1}{3}) \, \Gamma(\tfrac{2}{3}) = 2\pi 3^{-1/2}$, so

$$J_h(h) \doteqdot \Gamma(\tfrac{1}{3})/2^{2/3} 3^{1/6} h^{1/3} \pi = 0{\cdot}44731 \, h^{-1/3},$$

a result due to Cauchy, and given by Watson [114], Chapter VIII.

CHAPTER III

The Stokes Phenomenon

3.1 An elementary equation

If $w'' = h^2 w$, the W.K.B.J. solutions

$$(0, z) = e^{hz}, \qquad (z, 0) = e^{-hz}$$

are the exact solutions of the equation. In particular, consider

$$\cosh hz = \tfrac{1}{2}[(0, z) + (z, 0)] = \tfrac{1}{2}(e^{hz} + e^{-hz})$$
$$= \tfrac{1}{2}(e^{hx}e^{ihy} + e^{-hx}e^{-ihy}).$$

When $|z|$ is large, approximations may usefully be made.

If $\mathrm{Rl}\, z = x > 0$, $e^{hx} \gg e^{-hx}$, so $\cosh hz \doteqdot \tfrac{1}{2}e^{hz}$;

If $\mathrm{Rl}\, z = x < 0$, $e^{hx} \ll e^{-hx}$, so $\cosh hz \doteqdot \tfrac{1}{2}e^{-hz}$;

If $\mathrm{Rl}\, z = x = 0$, no approximation is valid.

Hence different approximations represent the same function in various domains of the z-plane.

In terms of $\arg z$,

$$\cosh hz \doteqdot \tfrac{1}{2}e^{hz} \qquad (-\tfrac{1}{2}\pi < \arg z < \tfrac{1}{2}\pi),$$
$$\cosh hz \doteqdot \tfrac{1}{2}e^{-hz} \qquad (\tfrac{1}{2}\pi < \arg z < \tfrac{2}{3}\pi),$$

while no approximation is possible on the lines $\arg z = -\tfrac{1}{2}\pi$ and $\tfrac{1}{2}\pi$. These lines, given by $\mathrm{Rl}\, z = 0$, are the anti-Stokes lines for the function. For $-\tfrac{1}{2}\pi < \arg z < \tfrac{1}{2}\pi$, e^{hz} is dominant and e^{-hz} is subdominant. This property reverses in character as e^{hz} or e^{-hz} is traced over an anti-Stokes line. Without approximations, we have

$$\cosh hz = \tfrac{1}{2}(0, z) + \tfrac{1}{2}(z, 0) \qquad (\arg z = -\tfrac{1}{2}\pi),$$
$$\cosh hz = \tfrac{1}{2}(0, z)_d + \tfrac{1}{2}(z, 0)_s \qquad (-\tfrac{1}{2}\pi < \arg z < \tfrac{1}{2}\pi),$$
$$\cosh hz = \tfrac{1}{2}(0, z) + \tfrac{1}{2}(z, 0) \qquad (\arg z = \tfrac{1}{2}\pi),$$
$$\cosh hz = \tfrac{1}{2}(0, z)_s + \tfrac{1}{2}(z, 0)_d \qquad (\tfrac{1}{2}\pi < \arg z < \tfrac{3}{2}\pi).$$

Very near an anti-Stokes line, both terms must be retained, since although $|e^{-hz}|$ is exponentially small when it is subdominant, yet as z approaches the anti-Stokes line $\mathrm{Rl}\, z = 0$, the value of $|e^{-hz}|$ increases quickly to unity. If we agree to allow an error $|e^{hz} O(1/h)|$ in $\cosh hz$, then as soon as $|e^{-hz}|$ is greater than this error it must be included. The boundary of the thin strip along the anti-Stokes line in which the subdominant term must be included is therefore specified by

$$e^{hx} O(1/h) = e^{-hx},$$

or

$$x = -\frac{1}{2h} \log O\!\left(\frac{1}{h}\right).$$

There is thus an element of doubt in the boundary of this strip around the anti-Stokes line, doubt caused by ignorance of the exact numerical value to be attached to $O(1/h)$. In this case of course we could be definite as to the value we require to be attached to $O(1/h)$, but in the W.K.B.J. solutions, it is not the province of the theory to examine the bounds that may be attached to the term $O(1/h)$.

3.2 The Airy equation

Although the *Airy equation* (or the *Stokes equation*)

$$w'' = zw \tag{3.1}$$

does not contain a large parameter h, the approximate method of solution developed in section 2.4 is still applicable, but with a different form necessary for the error estimate.

Following section 2.2, we choose

$$\xi = \int\limits_0^z z^{1/2} \, dz = \tfrac{2}{3} z^{3/2},$$

where a suitable branch cut is inserted with $\arg z^{1/2}$ chosen to vanish along the positive x-axis. Then if $w = z^{-1/4} X$, equation (2.5) becomes

$$X'' - X = -\frac{5}{36} \xi^{-2} X. \tag{3.2}$$

Equation (3.1) may of course be transformed directly into this form by using $w = z^{-1/4} X$, $\xi = \tfrac{2}{3} z^{3/2}$.

The integral equation (2.13) for X is

$$X(\xi) = A e^{\xi} + B e^{-\xi} - \frac{5}{72} \int_C^{\xi} (e^{\xi - t} - e^{-\xi + t}) t^{-2} X(t)\, dt.$$

For the subdominant approximation $e^{-\xi}$, let C be chosen to be $+\infty$ on the real ξ-axis. Choose the path of integration from C to P via Q

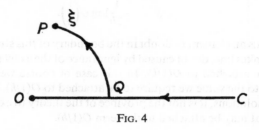

FIG. 4

as shown in Fig. 4, where $OP = OQ = |\xi|$ and PQ is an arc of a circle centre O. Then

$$\int_C^{\xi} \left| \frac{dt}{t^2} \right| = \int_C^{Q} + \int_Q^{\xi} = \left| \int_{\infty}^{|\xi|} \frac{dt}{t^2} \right| + \frac{\text{arc } PQ}{|\xi|^2}$$

$$(t \text{ real and positive})$$

$$= \frac{1}{|\xi|} + \frac{|\xi| \angle POQ}{|\xi|^2} = O\!\left(\frac{1}{\xi}\right).$$

Hence the subdominant solution is given by

$$X(\xi) = e^{-\xi}[1 + O(\xi^{-1})]$$

valid for

$$-\tfrac{3}{2}\pi < \arg \xi < \tfrac{3}{2}\pi,$$

that is, at all points ξ attained from C by paths along which $\mathrm{Rl}\, t$ decreases. The definite choice of path CQP was made in order to avoid artificial zig-zag paths along which the above integral of the modulus of the integrand may not even converge.

For $\arg \xi > \pi$ (or $< -\pi$), this method of estimating the error is rather rough, since from Fig. 5 we see that

$$\int_C^Q + \int_Q^{Q'} + \int_{Q'}^Q = O\left(\frac{1}{OQ}\right) = O\left(\frac{1}{R1\xi}\right),$$

contributions that arise from all three integrals. This difficulty did not arise in section 2.4, since the estimate was in terms of $O(1/h)$.

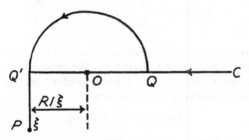

FIG. 5

It follows that the solution is valid only well away from the transition point O, namely when $|z| \gg 1$.

In this simple case, the exact form of the first correction term may easily be calculated. The correction term is

$$-\frac{5}{72} \int_\infty^\xi (e^{\xi - 2t} - e^{-\xi}) \frac{dt}{t^2}.$$

The first exponential, when integrated by parts, yields terms of the form $O(1/\xi^2)$, but the second yields

$$-\frac{5}{72} \left[\frac{e^{-\xi}}{t}\right]_\infty^\xi = -\frac{5}{72} \cdot \frac{e^{-\xi}}{\xi}.$$

In terms of z, the subdominant solution is

$$w = z^{-1/4} e^{-\xi} \left[1 - \frac{5}{72\xi} + O\left(\frac{1}{\xi^2}\right) \right] = z^{-1/4} \exp\left(-\tfrac{2}{3}z^{3/2}\right) \times$$

$$\times \left[1 - \frac{5}{48z^{3/2}} - O\left(\frac{1}{z^3}\right) \right], \quad \vdots \qquad (3.3)$$

these being the first two terms in the asymptotic development of the solution (see Jeffreys and Jeffreys [68]).

For the dominant solution e^{ξ}, the first correction term is given by

$$X(\xi) = e^{\xi} - \frac{5}{72} \int_{C}^{\xi} (e^{\xi} - e^{-\xi + 2t}) \frac{dt}{t^2}.$$

The fact that we are not finding an estimate of the form $O(1/h)$ necessitates a slight change of procedure here. We choose $C = +i\infty$ on the anti-Stokes line forming part of the boundary of the domain of validity of the solution. Then, as before,

$$X(\xi) = e^{\xi}[1 + O(1/\xi)],$$

valid for $-\tfrac{1}{2}\pi < \arg \xi < \tfrac{1}{2}\pi$, the estimate being rather rough for $\arg \xi < 0$. In terms of z,

$$w = z^{-1/4} \exp\left(\tfrac{2}{3}z^{3/2}\right) \left[1 + O\left(\frac{1}{z^{3/2}}\right) \right].$$

These results and ranges of validity may also be obtained by the method of steepest descents based on contour-integral representations of the solutions of equation (3.1); see Jeffreys and Jeffreys [68] and Budden [24].

3.3 Interpretation in the complex z-plane
The W.K.B.J. solutions of the differential equation

$$w'' = zw$$

are made up of the two expressions

$$(0, z) = z^{-1/4} \exp\left(\tfrac{2}{3}z^{3/2}\right), \qquad (z, 0) = z^{-1/4} \exp\left(-\tfrac{2}{3}z^{3/2}\right).$$

Anti-Stokes lines exist when $\text{Rl}\, z^{3/2} = 0$, namely when $\arg z = \pm\frac{1}{3}\pi, \pi$, and so on. The expression $(0,z)_d$ is dominant in the domain $-\frac{1}{3}\pi < \arg z < \frac{1}{3}\pi$, but subdominant, and therefore written as $(0,z)_s$, when $-\pi < \arg z < -\frac{1}{3}\pi$, $\frac{1}{3}\pi < \arg z < \pi$. Conversely, $(z,0)_s$ is subdominant when $-\frac{1}{3}\pi < \arg z < \frac{1}{3}\pi$ but dominant in the two neighbouring domains.

The branch cut from $z = 0$ may be inserted anywhere, provided that in any domain the forms of a particular expression are adjusted to maintain the required value. If $(0,z)_d$ is considered just on the negative side (obtained by a small clockwise rotation) of a branch cut at $\arg z = \delta$, then if $z = re^{i\delta}$,

$$(0,z)_d = (re^{i\delta})^{-1/4}\exp\left[-(re^{i\delta})^{3/2}\right]$$

$$= r^{-1/4}e^{-i\delta/4}\exp(\tfrac{2}{3}e^{3i\delta/2}r^{3/2}).$$

Just on the positive side of the cut, 2π must be subtracted from $\arg z$, so $z = re^{i\delta - 2\pi i}$, and now the form of the solution must be changed to $-i(z,0)_d$, since this equals

$$-i(re^{i\delta - 2\pi i})^{-1/4}\exp\left[-\tfrac{2}{3}(re^{i\delta - 2\pi i})^{3/2}\right]$$

$$= r^{-1/4}e^{-i\delta/4}\exp(\tfrac{2}{3}e^{3i\delta/2}r^{3/2})$$

as before. We therefore have rules for tracing expressions over the branch cut in the positive sense

$$\left.\begin{array}{l}(0,z) \to -i(z,0), \\ (z,0) \to -i(0,z),\end{array}\right\} \tag{3.4}$$

the property of dominancy or subdominancy being preserved in the process.

If such expressions are traced over the cut in a negative sense, the $-i$ is replaced by $+i$.

The implied error estimate $O(1/z^{3/2})$ is of course maintained; this rule merely changes the *form* of a solution, not its *value*.

Similarly, simple rules may be formulated for tracing expressions (which need not necessarily be solutions) across anti-Stokes lines in either sense. The property of dominancy or subdominancy is changed, yielding

$$\left.\begin{array}{r}(0, z)_s \rightarrow (0, z)_d, \\ (0, z)_d \rightarrow (0, z)_s, \\ (z, 0)_s \rightarrow (z, 0)_d, \\ (z, 0)_d \rightarrow (z, 0)_s. \end{array}\right\} \tag{3.5}$$

Consider now the three anti-Stokes lines OA, OB, OD with a branch cut OC. We may choose three solutions that are subdominant in the three domains 1, 2, 3 and 4 respectively, their validity extending over the bounding anti-Stokes lines up to but not including the neighbouring anti-Stokes lines.

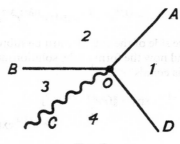

FIG. 6

Using rules (3.4) and (3.5), we find that these solutions are

Solution (i)

   1: $(z, 0)_s$,   2: $(z, 0)_d$,   4: $(z, 0)_d$,   3: $i(0, z)_d$,

     not determined on OB;

Solution (ii)

   2: $(0, z)_s$,   1: $(0, z)_d$,   3: $(0, z)_d$,   4: $-i(0, z)_d$,

     not determined on OD;

Solution (iii),

   3: $(z, 0)_s$,   4: $-i(0, z)_s$,   2: $(z, 0)_d$,   1: $-i(0, z)_d$,

     not determined on OA.

The first solution must be a linear combination of solutions (ii) and (iii), since the second-order Airy equation can only have two independent solutions; moreover, solutions (ii) and (iii) are obviously independent on account of the differing forms in, say, domain 4. We obtain, using A and B for the constants in the linear combination,

$$1: (z, 0)_s = A(0, z)_d - Bi(0, z)_d,$$
$$2: (z, 0)_d = A(0, z)_s + B(z, 0)_d,$$
$$\begin{cases} 3: i(0, z)_d = A(0, z)_d + B(z, 0)_s, \\ 4: (z, 0)_d = -Ai(0, z)_d - Bi(0, z)_s, \end{cases}$$

where the relations in domains 3 and 4 are of course identical.

From 1, the total dominant coefficient must vanish in order to yield the subdominant expression on the left, so $A - Bi = 0$.

From 2, the dominant coefficients must be identical on both sides of the equation, so $B = 1$.

From 3, the dominant coefficients must be identical, so $A = i$. Three consistent results are therefore produced.

This enables us to find the form of solution (i) on OB, where it is not as yet determined. For the solution is expressed in terms of solutions (ii) and (iii), both valid on OB. We have

$$\text{solution (i) on } OB = A(0, z) + B(z, 0)$$
$$= i(0, z) + (z, 0),$$

both expressions being neutral in magnitude. Hence the solution on OB is merely the sum of the terms that are dominant on each side of OB, namely $(z, 0)_d$ in 2 and $i(0, z)_d$ in 3.

It should be stressed that this representation for all arg z provides no information as to the value of the solution near or at $z = 0$.

3.4 The Stokes constants

The determination just given of the approximate solution for all arg z may be reduced to a series of simple rules by the introduction of *Stokes lines* and *Stokes constants*.

A *Stokes line* is defined by

$$\text{Im } z^{3/2} = 0,$$

namely the lines arg $z = 0$, $\pm \frac{2}{3}\pi$, and so on, bisecting the domains formed by the anti-Stokes lines. As a subdominant solution for fixed $|z|$ is traced across a Stokes line, it attains its position of maximum subdominancy, while a dominant expression attains its maximum dominancy.

Previously, between two anti-Stokes lines, only the dominant expression has been of relevance, while on an anti-Stokes line the solution has consisted of the two expressions dominant on each side of the line. We now change slightly the point of view, yet not so as to violate the uniformity of the approximation. An expression dominant on one side of an anti-Stokes line is extended up to the line, on which

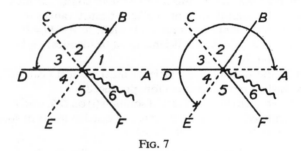

FIG. 7

it is needed, and over the line through the next $\frac{1}{3}\pi$ sector up to the neighbouring Stokes line. That is to say, in Fig. 7 a solution dominant in domains 2 and 3 is extended into domains 1 and 4 up to the Stokes lines OA and OE respectively, Stokes lines being represented by broken lines. In sectors 1 and 4, the expression is subdominant, so it is irrelevant by contrast with a dominant solution that may be present, since the magnitude of the term $(z, 0)_s$, say, is much smaller than the actual error in the dominant expression $(0, z)_d O(1/z^{3/2})$. In the domains 4 and 5 in which the expression is subdominant, its coefficient must be changed discontinuously in order to emerge on the next anti-Stokes line OF (on which it is needed) with the appropriate coefficient needed for the next domain 6 in which it is again dominant. This discontinuous change in no way violates the continuity of the approximate solution, since it is associated with a term much smaller in magnitude than the error allowed in the dominant term. If no dominant term is present in domains 4 and 5, no change in the sub-dominant coefficient can take place. This change suffered by the coefficient of the subdominant term is known as the *Stokes phenomenon*, after its discoverer Stokes [104].

This phenomenon may be dealt with quantitatively as follows. As a subdominant term is traced positively across a Stokes line, its new coefficient may be expressed in terms of the original coefficient and the coefficient of the dominant term. We write

New subdominant coefficient = old subdominant coefficient +

$$+ T \times \text{dominant coefficient}, \quad (3.6)$$

where T is the *Stokes constant* associated with the particular Stokes line.

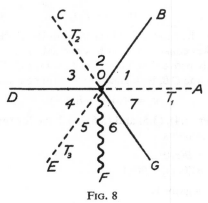

FIG. 8

The plane now appears as in Fig. 8, where T_1, T_2, T_3 denote the three Stokes constants.

For solution (i), subdominant in domains 1 and 7, we have previously had:

7, 1: $w = (z,0)_s,$
2, 3: $\quad (z,0)_d,$
OD: $\quad (z,0) + i(0,z),$
4, 5: $\quad i(0,z)_d,$
6: $\quad (z,0)_d.$

Under the new method of representation, this solution must be written as

7, 1: $\quad (z,0)_s,$

2: $\quad (z,0)_d,$

3: $\quad (z,0)_d + i(0,z)_s,$

4: $\quad (z,0)_s + i(0,z)_d,$

5: $\quad i(0,z)_d,$

6: $\quad (z,0)_d,$

where subdominant solutions have been inserted in domains 3 and 4 derived by extending over OD dominant solutions in domains 4 and 3 respectively. Clearly, on the Stokes line OC, $T_2 = i$, the coefficient of the subdominant term introduced, while on the Stokes line OE, (3.6) yields

$$0 = 1 + T_3 i,$$

so $T_3 = i$. When the Stokes line is crossed in the clockwise sense, the sign of the Stokes constants must be changed.

We may easily prove that $T_1 = T_2 = T_3 = i$ for all Stokes lines. On the anti-Stokes line OB, let a general solution be given by

$$w = A(0,z) + B(z,0).$$

Then using rules (3.4), (3.5) and (3.6), we have for the same solution in the various domains

2: $\quad A(0,z)_s + B(z,0)_d,$

3: $\quad (A + T_2 B)(0,z)_s + B(z,0)_d,$

positively, and negatively

1: $\quad A(0,z)_d + B(z,0)_s,$

7: $\quad A(0,z)_d + (B - T_1 A)(z,0)_s,$

6: $\quad A(0,z)_s + (B - T_1 A)(z,0)_d,$

5: $\quad iA(z, 0)_s + i(B - T_1 A)(0, z)_d,$

4: $\quad [iA - T_3 i(B - T_1 A)](z,0)_s + i(B - T_1 A)(0,z)_d.$

Comparing coefficients of $(z,0)$ and $(0,z)$ respectively on the anti-Stokes line OD, we obtain

$$B = iA - T_3 i(B - T_1 A),$$

$$A + T_2 B = i(B - T_1 A).$$

To be valid for all A and B, we may equate their respective coefficients, obtaining

$$1 = -T_3 i, \qquad 0 = i + T_3 T_1 i, \qquad 1 = -iT_1, \qquad T_2 = i,$$

yielding immediately $T_1 = T_2 = T_3 = i$ uniquely.

More complicated Stokes constants have been produced by the author in a series of papers [51, 52, 53] for the nth order differential equation

$$w^{(n)} = (-1)^n z^m w.$$

As a dominant term is traced across an anti-Stokes line, becoming subdominant in the process. it must be retained in the vicinity of the anti-Stokes line until its magnitude becomes smaller than the error allowed in the dominant term. Until this occurs, it is necessary to represent the solution as the sum of both terms, the subdominant term not being rejected. It is however difficult if not impossible to decide the boundaries of the region within which the subdominant term is retained. This difficulty arises on account of the use of a vague error term of the form $O(1/z^{3/2})$. Admittedly in this case something more definite could be said, since the numerical factor in this error term has been found, but in the general case this is not possible. This vagueness must be accepted as one of the inherent weaknesses of the phase-integral method.

Stokes [104] has shown that if the whole descending asymptotic series is appended to the W.K.B.J. approximation, and if it is desired to obtain the maximum accuracy in the summation of the descending asymptotic (divergent) series, then the change in the coefficient *must take place on the Stokes line itself*, the sum of both terms being retained throughout the domain between the anti-Stokes and the Stokes lines. But descending series are not the province of phase-integral methods, which are only concerned with the first term of the asymptotic expansions. It follows that the change in coefficient may be made anywhere between the somewhat vague boundaries just described outside of which the subdominant term is not required. It is only to be definite that the change is made on the Stokes line.

3.5 Standard solutions of the Airy equation

Stokes [104] represented the magnitudes of the two expressions $(0, z)$ and $(z, 0)$ for all $\arg z$ on a circular diagram, but we believe clarity is gained by representing $\arg z$ along a horizontal line AA' as in Fig. 9. BB' is taken to represent the neutral magnitude, CC' maximum dominancy and AA' maximum subdominancy. The expression $(0, z)$, dominant when $\arg z = 0$, is represented by the line $\dots QNS\dots$, and $(z, 0)$ subdominant when $\arg z = 0$, by the lines $\dots MRP\dots$. The solution subdominant when $\arg z = 0$ is represented by the heavy lines, the breaks occurring at N and P denoting the Stokes phenomenon.

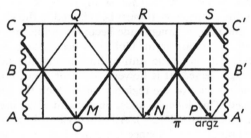

FIG. 9

Two standard solutions are chosen for the Airy equation as follows:

Solution (i). The solution $\mathrm{Ai}\,(z)$, valid for all z, is chosen such that its asymptotic expression is subdominant when $\arg z = 0$, $|z| \gg 1$. We write

$$\mathrm{Ai}\,(x) = \tfrac{1}{2}\pi^{-1/2}(x, 0)_s = \tfrac{1}{2}\pi^{-1/2} x^{-1/4} \exp\left(-\tfrac{2}{3} x^{3/2}\right) \quad (3.7)$$

for x real, large and positive, where the error term is implicitly implied. The reasons for the choice of the factor $\tfrac{1}{2}\pi^{-1/2}$ will be considered in the Appendix. Using Fig. 8, we have

$$
\begin{aligned}
1: &\quad \mathrm{Ai}\,(z) = \tfrac{1}{2}\pi^{-1/2}(z, 0)_s, \\
2: &\quad \mathrm{Ai}\,(z) = \tfrac{1}{2}\pi^{-1/2}(z, 0)_d, \\
3: &\quad \mathrm{Ai}\,(z) = \tfrac{1}{2}\pi^{-1/2}(z, 0)_d + \tfrac{1}{2}\pi^{-1/2} i(0, z)_s, \\
OD: &\quad \mathrm{Ai}\,(x) = \tfrac{1}{2}\pi^{-1/2}(z, 0) + \tfrac{1}{2}\pi^{-1/2} i(0, z).
\end{aligned}
$$

On the negative real axis, $z = xe^{i\pi}$, so

$$\begin{aligned}
\text{Ai}(-x) &= \tfrac{1}{2}\pi^{-1/2}(xe^{i\pi})^{-1/4}\exp\left[-\tfrac{2}{3}(xe^{i\pi})^{3/2}\right] + \\
&\quad + \tfrac{1}{2}\pi^{-1/2}i(xe^{i\pi})^{-1/4}\exp\left[\tfrac{2}{3}(xe^{i\pi})^{3/2}\right] \\
&= \tfrac{1}{2}\pi^{-1/2}x^{-1/4}[\exp(\tfrac{2}{3}ix^{3/2} - \tfrac{1}{4}\pi i) + \\
&\quad + i\exp(-\tfrac{2}{3}ix^{3/2} - \tfrac{1}{4}\pi i)] \\
&= \pi^{-1/2}x^{-1/4}\cos(\tfrac{2}{3}x^{3/2} - \tfrac{1}{4}\pi),
\end{aligned} \tag{3.8}$$

presenting Ai (x) as an evanescent real function on the positive real axis and as an oscillatory decreasing real function on the negative real axis.

Solution (ii). Bi (z) is defined to be dominant when arg $z = 0$, but by itself this does not define the function uniquely. Unique definitions only exist along anti-Stokes lines or along a Stokes line on which the solution is subdominant. On the negative real axis, solution (ii) is defined to equal solution (i) but with a phase difference of $\tfrac{1}{2}\pi$, namely

$$\text{Bi}(-x) = \pi^{-1/2}x^{-1/4}\cos(\tfrac{2}{3}x^{3/2} + \tfrac{1}{4}\pi) \tag{3.9}$$

$$= \tfrac{1}{2}\pi^{-1/2}i(z,0) + \tfrac{1}{2}\pi^{-1/2}(0,z)$$

on OD when $z = xe^{i\pi}$. Then we have

$$3: \ \text{Bi}(z) = \tfrac{1}{2}\pi^{-1/2}i(z,0)_d + \tfrac{1}{2}\pi^{-1/2}(0,z)_s,$$
$$2: \ \text{Bi}(z) = \tfrac{1}{2}\pi^{-1/2}i(z,0)_d + (\tfrac{1}{2}\pi^{-1/2} - i.\tfrac{1}{2}\pi^{-1/2}i)(0,z)_s,$$
$$1: \ \text{Bi}(z) = \tfrac{1}{2}\pi^{-1/2}i(z,0)_s + \pi^{-1/2}(0,z)_d.$$

Hence

$$\text{Bi}(x) = \pi^{-1/2}(0,x)_d = \pi^{-1/2}x^{-1/4}\exp(\tfrac{2}{3}x^{3/2}) \tag{3.10}$$

along the positive real axis.

3.6 A glance at Bessel's equation
In example 2, section 2.7, $u = wz^{-1/2}$, where

$$w'' + (1 - H^2 z^{-2})w = 0.$$

When $|z|$ is large, the W.K.B.J. solutions merely reduce to $e^{\pm iz}$, yielding the W.K.B.J. approximations of Bessel's equation in the form

$z^{-1/2}e^{\pm iz}$. Anti-Stokes lines exist when $\mathrm{Rl}\,(iz) = 0$, namely along the real axis, while Stokes lines exist along the imaginary axis.

Now if a series expansion

$$u = z^c(1 + a_1 z + a_2 z^2 + \ldots)$$

is substituted into Bessel's equation of order ν

$$u'' + z^{-1} u' + (1 - \nu^2 z^{-2}) u = 0,$$

we obtain formally

$$
\begin{aligned}
&c(c-1)z^{c-2} + a_1(c+1)\,cz^{c-1} + a_2(c+2)(c+1)\,z^c + \ldots \\
&+ \quad cz^{c-2} \quad + a_1(c+1)z^{c-1} + \quad a_2(c+2)\,z^c \quad + \ldots \\
&\qquad\qquad\qquad\qquad\qquad + \quad\quad z^c \quad\quad + \ldots \\
&- \quad \nu^2 z^{c-2} \quad - \quad \nu^2 a_1 z^{c-1} \quad - \quad \nu^2 a_2 z^c \quad\quad -\ldots = 0.
\end{aligned}
$$

Equating coefficients of z^{c-2}, z^{c-1}, z^c, ... to zero, we obtain

$$c(c-1) + c - \nu^2 = 0,$$

$$a_1(c+1)\,c + a_1(c+1) - \nu^2 a_1 = 0,$$

$$a_2(c+2)(c+1) + a_2(c+2) + 1 - \nu^2 a_2 = 0,$$

$$a_3(c+3)(c+2) + a_3(c+3) + a_1 - \nu^2 a_2 = 0.$$

The first equation gives $c = \pm\nu$, and the second equation gives $a_1 = 0$. The remaining equations determine $a_2, a_4, a_6, \ldots$ in terms of c, with $a_1 = a_3 = a_5 = \ldots = 0$. The general solution of Bessel's equation therefore takes the form

$$
\begin{aligned}
u &= Az^\nu[\text{power series (i) in } z^2] + \\
&\quad + Bz^{-\nu}[\text{power series (ii) in } z^2] \\
&= Az^\nu P_1(z^2) + Bz^{-\nu} P_2(z^2)
\end{aligned}
\tag{3.11}
$$

say, where the two series converge for all z. Modifications are necessary when ν is an integer.

When $|z|$ is large with $\arg z = 0$, consider a solution of the form

$$u = z^{-1/2} e^{iz}, \tag{3.12}$$

and we shall suppose that A and B are chosen so that solution (3.11) with $\arg z = 0$ equals solution (3.12) with $\arg z = 0$ on the anti-Stokes line $\arg z = 0$.

The solution is subdominant for $0 < \arg z < \pi$, so no change can take place on the Stokes line $\arg z = \frac{1}{2}\pi$ in the absence of a dominant term. After $\arg z = \pi$, the solution becomes dominant, and after the Stokes line $\arg z = \frac{3}{2}\pi$ with T as the Stokes constant the solution becomes

$$u = z^{-1/2}e^{iz} + Tz^{-1/2}e^{-iz} \tag{3.13}$$

both terms being neutral on the next anti-Stokes line $\arg z = 2\pi$.

Putting $z = x$ when $\arg z = 0$ in (3.11) and (3.12), we have

$$Ax^{\nu}P_1(x^2) + Bx^{-\nu}P_2(x^2) = x^{-1/2}e^{ix}.$$

Putting $z = xe^{i\pi}$ in (3.11) and (3.12), we have

$$A(xe^{i\pi})^{\nu}P_1(x^2) + B(xe^{i\pi})^{-\nu}P_2(x^2) = (xe^{i\pi})^{-1/2}e^{-ix}$$

the two power series retaining exactly the same form. Putting $z = xe^{2i\pi}$ in (3.11) and (3.13), we have

$$A(xe^{2i\pi})^{\nu}P_1(x^2) + B(xe^{2i\pi})^{-\nu}P_2(x^2)$$
$$= (xe^{2i\pi})^{-1/2}e^{ix} + T(xe^{2i\pi})^{-1/2}e^{-ix}.$$

The elimination of $Ax^{\nu}P_1(x^2)$ and $Bx^{-\nu}P_2(x^2)$ yields

$$\begin{vmatrix} 1 & 1 & x^{-1/2}e^{ix} \\ e^{i\pi\nu} & e^{-i\pi\nu} & -ix^{-1/2}e^{-ix} \\ e^{2i\pi\nu} & e^{-2i\pi\nu} & -x^{-1/2}e^{ix} - Tx^{-1/2}e^{-ix} \end{vmatrix} = 0.$$

Upon expansion, the factors $e^{\pm ix}$ cancel entirely, leaving

$$T = 2i\cos\pi\nu. \tag{3.14}$$

It can further be shown that this is the Stokes constant for all Stokes lines for Bessel's equation, provided of course that the basic W.K.B.J. solutions are $z^{-1/2}e^{\pm iz}$. If different constants of proportionality are inserted (as in the case of the two Hankel functions), T must be modified accordingly, but this modification is trivial.

3.7 The equation with two transition points

Consider the equation

$$w'' + (z^2 - a^2) w = 0,$$

for which exact single-valued solutions may be obtained; these are known as the *Weber parabolic cylinder functions* (see Whittaker and Watson [116]).

If a is real, $q = \sqrt{(z^2 - a^2)}$ is real along the real z-axis when $|x| > 0$, and purely imaginary between the two zeros of q at $x = \pm a$.

When $|z|$ is large, we have

$$i \int^z q^{1/2} \, dz = i \int^z \sqrt{(z^2 - a^2)} \, dz$$

$$\doteqdot i \int^z z(1 - \tfrac{1}{2}a^2 z^{-2}) \, dz \qquad |z| \gg |a|$$

$$= i(\tfrac{1}{2}z^2 - \tfrac{1}{2}a^2 \log z);$$

hence the two W.K.B.J. solutions may be taken to be

$$z^{-1/2} \exp\left[\pm i(\tfrac{1}{2}z^2 - \tfrac{1}{2}a^2 \log z) \right] \equiv z^{-(1/2) \pm ia^2/2} \exp\left(\pm \tfrac{1}{2} iz^2 \right).$$

The lower limit of integration, k say, is irrelevant; but to be definite regarding notation, we shall write

$$(k, z) \equiv z^{-(1/2) - ia^2/2} \exp\left(\tfrac{1}{2} iz^2 \right),$$

$$(z, k) \equiv z^{-(1/2) + ia^2/2} \exp\left(-\tfrac{1}{2} iz^2 \right). \tag{3.15}$$

Anti-Stokes lines occur when $\mathrm{Rl}\,(iz^2) = 0$, namely along the positive and negative real and imaginary axes. Stokes lines occur when $\mathrm{Im}\,(iz^2) = 0$, along lines bisecting the axes. The Stokes phenomenon is not easy to deal with in this case; a fuller discussion is given in the Appendix, but here we shall consider only the important results.

If a branch cut is inserted at $\arg z = \delta$, and if $z = re^{i\delta}$ just before the cut, then

$$(k, z) = (re^{i\delta})^{-(1/2) - ia^2/2} \exp\left[\tfrac{1}{2} i(re^{i\delta})^2 \right],$$

but just after the branch cut we must have $z = re^{i\delta - 2\pi i}$, so

$$(k, z) = (re^{i\delta})^{-(1/2) - ia^2/2}(e^{-2\pi i})^{-(1/2) - ia^2/2} \exp\left[\tfrac{1}{2}i(re^{i\delta})^2\right].$$

That is to say,

$$(k, z) \text{ before the cut} \equiv -e^{\pi a^2} \times (k, z) \text{ after the cut,}$$

and similarly,

$$(z, k) \text{ before the cut} \equiv -e^{-\pi a^2} \times (z, k) \text{ after the cut.}$$

Hence we have the rules

$$\left.\begin{array}{l} (k, z) \rightarrow -e^{\pi a^2}(k, z), \\ (z, k) \rightarrow -e^{-\pi a^2}(z, k), \end{array}\right\} \qquad (3.16)$$

for a positive crossing of the branch cut, while for a negative crossing the sign in the exponent of the exponential is merely changed.

Fig. 10 shows the magnitude of the W.K.B.J. expressions when a branch cut is inserted just below the negative real axis. S, T, U, V

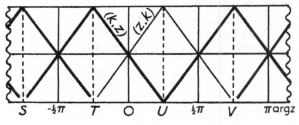

FIG. 10

denote the four Stokes constants at the four Stokes lines, and the heavy lines represent that solution subdominant when $\arg z = \tfrac{1}{4}\pi$; three changes of coefficients evidently take place for this function.

Consider now a given solution $w = A(k, z) + B(z, k)$ on the anti-Stokes line $\arg z = 0$. To be more precise, we shall consider $\arg z = 2p\pi$, where p is an integer, since the value of p enters the values of the Stokes constants.

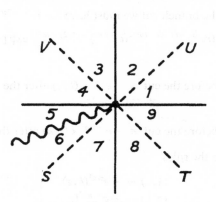

FIG. 11

Tracing the solution round positively, we have

1: $A(k,z)_s + B(z,k)_d$,
2: $(A+UB)(k,z)_s + B(z,k)_d$,
3: $(A+UB)(k,z)_d + B(z,k)_s$,
4: $(A+UB)(k,z)_d + (B+VA+VUB)(z,k)_s$,
5: $(A+UB)(k,z)_s + (B+VA+VUB)(z,k)_d$,

and negatively,

9: $A(k,z)_d + B(z,k)_s$,
8: $A(k,z)_d + (B-TA)(z,k)_s$,
7: $A(k,z)_s + (B-TA)(z,k)_d$,
6: $(A-SB+STA)(k,z)_s + (B-TA)(z,k)_d$,
5: $-(A-SB+STA)e^{-\pi a^2}(k,z)_s - (B-TA)e^{\pi a^2}(z,k)_d$,

using the rules (3.16) in the negative sense. Equating coefficients of (k,z) and of (z,k) respectively, we obtain

$$A + UB = -e^{-\pi a^2}(A - SB + STA),$$

$$B + VA + VUB = -e^{\pi a^2}(B - TA),$$

and finally the respective equating of the coefficients of A and B yields

$$1 = -e^{-\pi a^2}(1 + ST),$$

$$U = e^{-\pi a^2}S,$$

$$V = e^{\pi a^2}T,$$

$$1 + VU = -e^{\pi a^2}.$$

Unfortunately (unlike the case of the Airy integral) these four equations do not provide the values of the four constants, since the last three equations may be rearranged to yield the first. In terms of U,

$$S = e^{\pi a^2}U,$$

$$T = -(e^{\pi a^2} + 1)/S = -(2e^{-\pi a^2/2}\cosh \tfrac{1}{2}\pi a^2)/U,$$

$$V = e^{\pi a^2}T = -(2e^{\pi a^2/2}\cosh \tfrac{1}{2}\pi a^2)/U.$$

Now it can be proved that

$$U = i\sqrt{(2\pi)}\, 2^{-ia^2/2}\, e^{-2p\pi a^2 - \pi a^2/4}/\Gamma(\tfrac{1}{2} - \tfrac{1}{2}ia^2), \qquad (3.17)$$

where Γ denotes the gamma function. Then

$$\left.\begin{aligned}
S &= i\sqrt{(2\pi)}\, 2^{-ia^2/2}\, e^{-2p\pi a^2 + 3\pi a^2/4}/\Gamma(\tfrac{1}{2} - \tfrac{1}{2}ia^2), \\
T &= i\sqrt{(2\pi)}\, 2^{ia^2/2}\, e^{2p\pi a^2 - \pi a^2/4}/\Gamma(\tfrac{1}{2} + \tfrac{1}{2}ia^2), \\
V &= i\sqrt{(2\pi)}\, 2^{ia^2/2}\, e^{2p\pi a^2 + 3\pi a^2/4}/\Gamma(\tfrac{1}{2} + \tfrac{1}{2}ia^2).
\end{aligned}\right\} \qquad (3.18)$$

In deriving these values from U, we have used the relation associated with the gamma function

$$\Gamma(\tfrac{1}{2} - \tfrac{1}{2}ia^2)\, \Gamma(\tfrac{1}{2} + \tfrac{1}{2}ia^2) = \frac{\pi}{\sin \pi(\tfrac{1}{2} - \tfrac{1}{2}ia^2)} = \frac{\pi}{\cos \tfrac{1}{2}\pi ia^2} = \frac{\pi}{\cosh \tfrac{1}{2}\pi a^2}.$$

These values of the four Stokes constants enable asymptotic expressions for the solutions of the original equation to be written down for all $\arg z$. The values must be used with care. For example, if the branch cut is placed in a different sector, the constants must be changed accordingly. Generally, the Stokes constant for the Stokes line $\arg z = \frac{1}{2} r\pi - \frac{1}{4}\pi$ is

$$i\sqrt{(2\pi)}\, 2^{-ia^2/2}\, e^{\pi a^2(1/2-r)/2} / \Gamma(\tfrac{1}{2} - \tfrac{1}{2}ia^2)$$

if r is odd, and

$$i\sqrt{(2\pi)}\, 2^{ia^2/2}\, e^{\pi a^2(r-1/2)/2} / \Gamma(\tfrac{1}{2} + \tfrac{1}{2}ia^2)$$

if r is even.

CHAPTER IV

One Transition Point

4.1 The W.K.B.J. solutions

In Chapter II, we have seen that differential equations of the form

$$w'' + h^2 q(z, h) w = 0$$

often admit of approximate W.K.B.J. solutions in the form

$$(a, z) = q^{-1/4} \exp\left(ih \int_a^z q^{1/2} dz \right) \left[1 + O\left(\frac{1}{h}\right) \right],$$

$$(z, a) = q^{-1/4} \exp\left(-ih \int_a^z q^{1/2} dz \right) \left[1 + O\left(\frac{1}{h}\right) \right],$$

in restricted domains of the complex z-plane.

We now consider how more complete solutions may be obtained, solutions that do not suffer from this restriction on their domains of validity. To this end, we consider the simplest yet most usual case of one transition point of *order unity*. That is to say, q possesses a zero at the point $z = z_0$ (chosen to be zero for convenience) such that in its neighbourhood q exists as a convergent power series of the form

$$q = a_1 z + a_2 z^2 + a_3 z^3 + \ldots, \tag{4.1}$$

where $a_1 \neq 0$. If z^n is the first non-vanishing term in the expansion, the transition point is said to be of *order n*.

The expressions (a, z) and (z, a) are valid in domains excluding $z = 0$ and bounded by anti-Stokes lines defined by

$$\mathrm{Rl}\, i \int_0^z q^{1/2} dz = 0;$$

three such lines radiate outwards from a transition point of order 1.

These approximations may fail if $|z|$ becomes too large. The presence of further zeros of q, and difficulties with the path of integration discussed in section 2.4 restrict the domains of validity for large $|z|$, but these questions are too involved to be considered here; the subject has been dealt with by the author elsewhere [56].

Following the methods adopted in Chapter III, we shall now define Stokes lines lying between the anti-Stokes lines by the equation

$$\operatorname{Im} i \int_0^z q^{1/2} dz = 0.$$

The neighbourhood of the point $z = 0$ thus contains seven lines radiating outwards from $z = 0$: three anti-Stokes lines (full lines), three Stokes lines (broken lines) and a branch cut (a wavy line), as in Fig. 12.

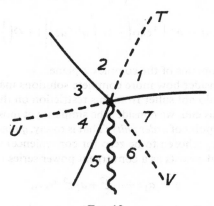

Fig. 12

In section 2.2 equation (2.10), we have produced an approximation that applies throughout the plane depicted in Fig. 12, though this solution will break down further outwards towards further zeros of q. If $z \neq 0$, we may use the asymptotic expressions of the Airy function (section 3.3), and it may easily be shown that these asymptotic expressions are merely proportional to the W.K.B.J. solutions of the equation.

Following now the procedure adopted in section 3.4, we extend the W.K.B.J. expressions over their present anti-Stoke line boundaries up to the neighbouring Stokes lines, and Stokes constants are introduced for each of the three Stokes lines as in Fig. 12. This extension in no way invalidates the error terms of the form $O(1/h)$, since the subdominant terms thus introduced are irrelevant by comparison with the error in the dominant terms.

When a solution is traced across the branch cut, we preserve continuity in the positive sense by changing

$$(0, z) \to -i(z, 0),$$

$$(z, 0) \to -i(0, z) \tag{4.2}$$

as in section 3.3

If $w = A(0, z) + B(z, 0)$ is an assumed solution along the anti-Stokes line between domains 1 and 7, and if $(0, z)$ is subdominant in 1, then the solution is

7: $A(0, z)_d + B(z, 0)_s$,
1: $A(0, z)_s + B(z, 0)_d$,
2: $(A + TB)(0, z)_s + B(z, 0)_d$,
3: $(A + TB)(0, z)_d + B(z, 0)_s$,
4: $(A + TB)(0, z)_d + (B + UA + UTB)(x, 0)_s$,
5: $(A + TB)(0, z)_s + (B + UA + UTB)(z, 0)_d$,
6: $-i(A + TB)(z, 0)_s - i(B + UA + UTB)(0, z)_d$,
7: $-i(A + TB + VB + VUA + VUTB)(z, 0)_s -$
   $-i(B + UA + UTB)(0, z)_d$.

Comparing coefficients in domain 7, we find that

$$1 = -iU, \qquad\qquad 0 = -i(1 + UT),$$

$$0 = -i(1 + VU), \qquad 1 = -i(T + V + VUT),$$

yielding $T = U = V = i$. This result should have been obvious; these are merely the Stokes constants for the Airy function occurring in the solution (2.10) valid throughout the domain under consideration. The method given here is a systematic development of that given by Furry [37].

4.2 The representation of waves

Many problems associated with the propagation of waves in a homogeneous or inhomogeneous medium are governed by a partial differential equation of the form

$$\frac{\partial^2 W}{\partial x^2} = \frac{q(x)}{c^2} \cdot \frac{\partial^2 W}{\partial t^2},$$

where t is the time and where $q(x)$ is a property of the medium through which the waves are being transmitted; c is the velocity of the waves when $q(x) \equiv 1$.

For a particular frequency, the time factor may be represented by $e^{i\omega t}$, and if $W = w(x)e^{i\omega t}$, we obtain

$$\frac{d^2 w}{dx^2} + q(x)(\omega^2/c^2)\, w = 0,$$

or

$$\frac{d^2 w}{dx^2} + h^2 q(x)\, w = 0, \tag{4.3}$$

where $h = \omega/c$. Sometimes the equation may be formulated *ab initio* for a particular frequency, and in this case $q(x)$ must be replaced by $q(x,h)$ if some property of the medium is frequency-dependent. The propagation of electromagnetic waves through an isotropic ionized medium provides an example of this.

If $q(x)$ is real, positive and constant, the medium will transmit waves with phase velocity $c/q^{1/2}$, while if q is real, constant and negative the waves will be evanescent with infinite phase velocity. In the former case, the solution for w will contain complex exponentials, in the latter case real exponentials. In the former case, the medium is said to be *underdense*, and in the latter case *overdense*.

Generally, if q is complex and constant, solutions of equation (4.3) are

$$w = \exp(\pm ihq^{1/2}x) = \exp[\pm ih(\alpha - i\beta)x],$$

where $q \equiv (\alpha - i\beta)^2$. Hence

$$W = e^{-ih\alpha(x - ct/\alpha)} e^{-h\beta x}, \tag{4.4}$$

or

$$W = e^{ih\alpha(x + ct/\alpha)} e^{h\beta x}. \tag{4.5}$$

Solution (4.4) represents a wave travelling in the positive x-direction with phase velocity c/α and attenuated by the factor $e^{-h\beta x}$; (4.5) represents a wave travelling in the negative x-direction with the attenuation governed by the factor $e^{h\beta x}$. It is essential that α and β should both be positive, in order that attenuation should take place in the direction of propagation. Hence $q^{1/2} \equiv \alpha - i\beta$ must be in the fourth quadrant (or in the second quadrant if the branch is changed).

If q is now a function of x, the W.K.B.J. solutions represent waves propagated independently if an anti-Stokes line lies along the real axis. If the W.K.B.J. solutions do not exist (for example, if h is not large enough) for some range of x, then individual independent waves propagated to the right and to the left along the x-axis cannot be defined. Along the real axis, we have

$$W = we^{i\omega t} = q^{-1/4}\exp\left(\pm ih \int^{x} q^{1/2}\,dx\right)e^{i\omega t}.$$

Usually it is necessary to integrate through the complex z-plane, retaining the upper limit as $\mathrm{Rl}\,z \equiv x$ for physical reality. Then

$$W = q^{-1/4}\exp\left(\pm ih \int_{a}^{x} q^{1/2}\,dz\right)e^{i\omega t}. \tag{4.6}$$

Generally, the integral contains both real and imaginary parts, so both the argument (phase) and modulus (amplitude) of W vary with x. Taken together, the integral forms a generalized (complex) phase factor, yielding the name *Phase-Integral Methods*.

4.3 Direction of propagation of waves

If
$$ih \int_{a}^{z} q^{1/2}\,dz \equiv u(x,y) + iv(x,y)$$

when resolved into real and imaginary parts, the exponential part of W is

$$\exp\left[u(x,y) + iv(x,y) + i\omega t\right].$$

The factor $\exp[u(x,0)]$ constitutes an amplitude change along the real x-axis, and the factor $\exp[iv(x,0) \pm i\omega t]$ constitutes a phase change along the real x-axis.

The value of the phase at a nearby point $x + \delta x$ and at a slightly later time $t + \delta t$ is

$$i\left(v(x,0) + \delta x \cdot \frac{\partial v(x,0)}{\partial x}\right) + i\omega t + i\omega\delta t$$

to the first order. This equals the original phase (but of course with change of amplitude) if

$$\delta x \frac{\partial v(x,0)}{\partial x} + \omega\delta t = 0,$$

yielding the local phase velocity

$$\frac{dx}{dt} = -\omega \frac{\partial v(x,0)}{\partial x}.$$

Hence if $\partial v(x,0)/\partial x$ is negative, propagation takes place in the positive sense along Ox, while if it is negative, propagation takes place in the negative sense. This rule is of course reversed when the second W.K.B.J. solution is used.

In particular, consider the case when there is an anti-Stokes line lying along part of the real axis; we require to formulate a rule giving the direction of propagation of the two waves (a,x) and (x,a).

As far as (a,x) is concerned, $u(x,0) = 0$ since the axis is an anti-Stokes line. Now $u + iv$ is a function of the complex variable z, so the *Riemann-Cauchy equations* hold, namely

$$\frac{\partial u}{\partial x} = \frac{\partial v}{\partial y}, \qquad \frac{\partial u}{\partial y} = -\frac{\partial v}{\partial x}.$$

Now if (a,x) represents a wave propagated to the right (namely along $+Ox$), we have seen that $\partial v(x,0)/\partial x < 0$; hence $\partial u(x,0)/\partial y > 0$. Then near the real axis, we have

$$u(x, \delta y) = u(x,0) + \delta y \frac{\partial u(x,0)}{\partial y},$$

using the first two terms of the Taylor expansion of u. Since $u(x,0) = 0$, it follows that $u(x, \delta y)$ becomes negative if $\delta y < 0$ below the real axis. In other words, (a, x) becomes subdominant.

Rule: Along an anti-Stokes line along the real axis, (a, x) represents a wave propagated to the right if it becomes subdominant below the axis. It represents a wave propagated to the left if it becomes dominant below the axis. The same results hold for the second solution (x, a).

This rule has been called Heading's rule†, since it appears to have been first formulated by the author [50]. It should be pointed out that the rule is reversed if the time factor $e^{i\omega t}$ is replaced by $e^{-i\omega t}$.

4.4 Conservation of energy flow

If $q(x)$ is real along the real axis Ox, and if a star denotes the complex conjugate, we have

$$w'' + h^2 q^2 w = 0$$

and

$$w^{*''} + h^2 q^2 w^* = 0$$

along the real axis. Multiplying the first equation by w^* and the second by w and subtracting, we obtain

$$w'' w^* - w^{*''} w = 0,$$

or

$$\frac{d}{dx}(w' w^* - w^{*'} w) = 0,$$

implying that

$$\mathrm{Im}\,(w' w^*) = \text{constant}, \tag{4.7}$$

or

$$\mathrm{Im}\,(W' W^*) = \text{constant}.$$

This conservation equation may be related to the conservation of the flow of energy, provided of course that W has the right physical dimensions for $W' W^*$ to denote energy flow. The conservation is understood in the sense of a time average. If W is an electric field, and W' the magnetic field perpendicular to W in the wave front, then $W' W^*$ is proportional to the complex Poynting vector.

† The author would not seek to perpetuate his name in this connection, but this name has been advocated by others; see Budden [24], page 442.

The medium is described as *loss-free* under these circumstances. If $q(x)$ is complex along Ox, no such integral as (4.7) exists along the real axis.

If $q(x) > 0$, and if

$$w = A(a, x) + B(x, a) \qquad (A, B \text{ real})$$

where a lies in the range for which $q(x) > 0$, then

$$w^* = A(x, a) + B(a, x)$$

since only the i in the exponent of the exponential is changed in sign. Also

$$w' = ihAq^{1/2}(a, x) - ihBq^{1/2}(x, a)$$

where only the exponential is differentiated to obtain the required accuracy. Then

$$w' w^* = ih[A(x, a) + B(a, x)][Aq^{1/2}(a, x) - Bq^{1/2}(x, a)]$$
$$= ih(A^2 - B^2),$$

a pure imaginary number. Constant transmission of energy therefore takes place provided $A^2 \neq B^2$. If $A = \pm B$,

$$w \propto (a, x) + (x, a) \qquad \text{or} \qquad (a, x) - (x, a)$$

$$= 2q^{-1/4} \cos\left(h \int_a^x q^{1/2} \, dx\right) \quad \text{or } 2iq^{-1/4} \sin\left(h \int_a^x q^{1/2} \, dx\right),$$

representing standing waves for which no energy is transmitted.

If $q(x) < 0$, we may write $q(x) = e^{i\pi} r(x)$ with $r(x) > 0$. Then if

$$w = A(a, x) + B(x, a),$$
$$w^* = Ai(a, x) + Bi(x, a),$$

since the exponential factor remains the same but the factor $q^{-1/4}$ introduces the additional i. Also

$$w' = ihAq^{1/2}(a, x) - ihBq^{1/2}(x, a)$$

as before, so

$$w' w^* = -h[A(a, x) + B(x\, a)][Aq^{1/2}(a,x) - Bq^{1/2}(x, a)]$$
$$= -h[A^2 q^{1/2}(a, x)^2 - B^2 q^{1/2}(x, a)^2]$$

in which expression the $q^{1/2}$ factor cancels entirely. Hence $w'w^*$ is real, so $\mathrm{Im}(w'w^*) = 0$, implying that there is no transmission of energy in an overdense medium when A and B are real.

If A and B are complex numbers, the above argument is modified to allow for the complex conjugates of A and B, but the conclusions are the same (see, however, section 4.9).

4.5 Jeffreys' connection formulae

Jeffreys' [64] original connection formulae were derived with $q(x)$ real along the real x-axis and possessing a simple zero at $x = 0$.

Approximate solutions with error terms of the form $O(1/h)$ were investigated for the equation

$$w'' + h^2 q(x,h)\, w = 0,$$

such that the approximate asymptotic expressions (the W.K.B.J. solutions) were known for both $x > 0$ and $x < 0$. We shall follow these conditions here, though the more general Fig. 12 may be used to connect solutions together along any anti-Stokes line and the opposite Stokes line.

We shall postulate a Stokes line along Ox, $x > 0$ and an anti-Stokes line along Ox, $x < 0$, with a branch cut below the real axis, as in Fig. 13.

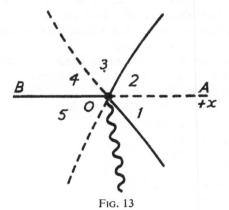

Fig. 13

Case (i), *subdominant solution on OA*. Let $w = (x, 0)$ be subdominant along OA. Then for the various domains we have

1, 2: $(z, 0)_s$,
3: $(z, 0)_d$,
4: $(z, 0)_d + i(0, z)_s$,
OB: $(x, 0) + i(0, x)$.

Following Jeffreys, we use the symbol $\leftrightarrow$ to indicate this connection, thus:

$$(x, 0) + i(0, x) \leftrightarrow (x, 0)_s. \qquad (4.8)$$

According to our work, we may proceed uniquely from the left to the right or from the right to the left, since either a subdominant or a neutral solution defines a unique solution to the equation. Jeffreys did not however use this symbol $\leftrightarrow$ with this meaning. He implied that there existed a solution for all x (including $x = 0$), such that its asymptotic expressions for large $|x|$ to the right and to the left took these respective forms. Jeffreys' notation has been often misunderstood, and as a result errors have been made in the literature. Perhaps it would have been better to have stated that there exists a solution $f(x)$ such that

$$(x, 0) + i(0, x) \leftarrow f(x) \rightarrow (x, 0)_s,$$

the arrows denoting the direction of implication from $f(x)$. Moreover, Jeffreys' original notation for a subdominant solution was such that later writers, not following his argument, have mistaken it for a dominant solution. Other writers have accused Jeffreys as being wrong, particularly in case (ii), but such accusations are unfounded when his work is examined in detail.

Case (ii), *dominant solution on OA*. A dominant solution along Ox does not define a unique solution. On the anti-Stokes line between 2 and 3 let a solution be

$$w = (0, z) + k(z, 0),$$

where k is any constant. Then we have

2: $(0, z)_d + k(z, 0)_s$,
1: $(0, z)_d + (k - i)(z, 0)_s$,
3: $(0, z)_s + k(z, 0)_d$,
4: $(1 + ik)(0, z)_s + k(z, 0)_d$,
OB: $(1 + ik)(0, x) + k(x, 0)$.

Along OA, $w = (0, x)_d$, the subdominant term being irrelevant, but along OB, the arbitrary constant k must remain. Jeffreys [64, 65] has produced two distinct versions of his second connection formula, with no explanation as to why one is more preferable than the other except a passing remark [67] to the effect that by using the second version 'no incorrect result has been reached'.

If we choose $k = \frac{1}{2}i$, this special value is such that the average of the two values of w on each side of OA (in domains 1 and 2) equals $(0, x)_d$. Then on OB,

$$w = \frac{1}{2}(0, x) + \frac{1}{2}i(x, 0).$$

Jeffreys would write

$$\tfrac{1}{2}(0, x) + \tfrac{1}{2}i(x, 0) \leftrightarrow (0, x)_d,$$

in the sense that a function $f(x)$ exists possessing the two asymptotic expressions indicated, and in no sense implying that we may pass from the right to the left uniquely. It would have been better had he written

$$\tfrac{1}{2}(0, x) + \tfrac{1}{2}i(x, 0) \leftarrow f(x) \rightarrow (0, x)_d,$$

the direction of the arrows denoting the sense of implication from $f(x)$. On the other hand, following recent writers, we use here the arrow $\rightarrow$ to denote the direction in which implication is unique:

$$\tfrac{1}{2}(0, x) + \tfrac{1}{2}i(x, 0) \rightarrow (0, x)_d. \tag{4.9}$$

This is, in effect, Jeffreys' second version.

His first version may be obtained by subtracting $\frac{1}{2}\sqrt{3}$ times the subdominant connection formula (4.8) from (4.9), yielding

$$\tfrac{1}{2}(1 - i\sqrt{3})(0, x) + \tfrac{1}{2}(i - \sqrt{3})(x, 0) \rightarrow (0, x)_d, \tag{4.10}$$

the dominant term being the only expression of relevance on the right-hand side. In other words, we choose $k = \frac{1}{2}(i - \sqrt{3})$ in case (ii).

Along OA, let $\arg q = -\pi$ and along OB, let $\arg q = 0$. Moreover, for $x > 0$, let

$$h \int_0^x q^{1/2} \, dx = -ih \int_0^x (-q)^{1/2} \, dx = -iM$$

say, where $M > 0$. Also for $x < 0$, let

$$h \int_0^x q^{1/2} dx = -L,$$

where $L > 0$ since the integral is taken from right to left.

Then (4.8) becomes

$$q^{-1/4} e^{iL} + iq^{-1/4} e^{-iL} \leftrightarrow [e^{-i\pi}(-q)]^{-1/4} e^{-M},$$

or upon multiplication by $e^{-i\pi/4}$,

$$q^{-1/4} e^{iL-(i\pi/4)} + q^{-1/4} e^{-iL+(i\pi/4)} \leftrightarrow (-q)^{-1/4} e^{-M},$$

that is,

$$2q^{-1/4} \sin(L + \tfrac{1}{4}\pi) \leftrightarrow (-q)^{-1/4} e^{-M}. \qquad (4.11)$$

Also (4.9) becomes

$$\tfrac{1}{2} q^{-1/4} e^{-iL} + \tfrac{1}{2} iq^{-1/4} e^{iL} \rightarrow [e^{-i\pi}(-q)]^{-1/4} e^{M},$$

yielding $\qquad q^{-1/4} \cos(L + \tfrac{1}{4}\pi) \rightarrow (-q)^{-1/4} e^{M}. \qquad (4.12)$

Finally (4.10) may be obtained by subtracting $\tfrac{1}{2}\sqrt{3}$ times formula (4.11) from (4.12), yielding

$$q^{-1/4} \cos(L + \tfrac{1}{4}\pi) - \sqrt{3} q^{-1/4} \sin(L + \tfrac{1}{4}\pi) \rightarrow (-q)^{-1/4} e^{M},$$

or $2q^{-1/4}[\cos \tfrac{1}{3}\pi \cos(L + \tfrac{1}{4}\pi) - \sin \tfrac{1}{3}\pi \sin(L + \tfrac{1}{4}\pi)] \rightarrow (-q)^{-1/4} e^{M}$,

or $\qquad 2q^{-1/4} \cos(L + \tfrac{7}{12}\pi) \rightarrow (-q)^{-1/4} e^{M}. \qquad (4.13)$

Formulae (4.8) and (4.9) are useful when working in the complex plane, but in this case they can be manufactured *ab initio* in each problem. Formulae (4.11) and (4.12) are useful when attention is confined to the real axis only, while formulae (4.10) and (4.13) are never used.

It should be pointed out that Jeffreys' papers should be read with care, since both his ideas and notation have developed over the years.

Further connection formulae may be produced by means of the Stokes constants. A wave transmitted to the right along BO becomes subdominant in domain 5 and dominant in 4 in keeping with the rule derived in section 4.3. Hence the solution is

OB: $(x,0)$,
4: $(z,0)_d$,
3: $(z,0)_d - i(0,z)_s$,
2: $(z,0)_s - i(0,z)_d$,

yielding the connection formula

$$(x,0) \rightarrow -i(0,x)_d.$$

Similarly, a wave transmitted to the left along OB yields

OB: $(0,x)$,
4: $(0,z)_s$,
3: $(0,z)_s$,
2: $(0,z)_d$,

yielding the connection formula

$$(0,x) \rightarrow (0,x)_d,$$

4.6 Reflection coefficients

It is now assumed that $q(z,h)$ has the following properties:

(i) $q(z,h) \rightarrow 1$ as $z \rightarrow -\infty$ along the real axis, though this constant may be replaced by other positive values,

(ii) $\mathrm{Rl}\,[h^2\,q(z,h)]$ becomes large and negative as $z \rightarrow \infty$ along the positive real axis. A complex transition point of order unity exists at $z = a$, at which point $q(z,h) = 0$; an anti-Stokes line emerges from $z = a$ and is asymptotic to the negative real axis, while the positive real axis lies within a domain formed by two anti-Stokes lines, as in Fig. 14.

A *phase-reference level* $z = b$ is chosen on the negative real axis where $q = 1$; then solutions (b,x) and (x,b) represent waves propagated along the negative real axis, being proportional to $e^{\pm ihx}$. Let the branch $q^{1/2}$ be chosen so that (b,x) represents a wave propagated to the left and (x,b) a wave to the right.

We shall define (x,b) to be the *incident wave*, with respect to the phase reference level b. Since the medium is overdense at $x = +\infty$, the required physical boundary condition is that w should be sub-dominant along $+Ox$, namely in domains 1 and 2. Let $(z,b)_s$ be the subdominant solution in 1 and 2. In tracing the solution round we must use (z,a) and not (z,b) since the Stokes constant 'i' applies

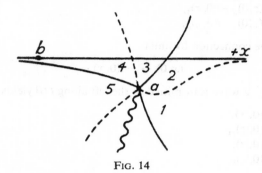

FIG. 14

when the phase-reference level is at the transition point. Then we have

$$1, 2: \quad (z,b)_s \equiv [a,b](z,a)_s,$$
$$3: \quad [a,b](z,a)_d,$$
$$4: \quad [a,b](z,a)_d + i[a,b](a,z)_s,$$
$$-Ox: \quad (x,b) + i[a,b]^2(b,x).$$

Now (x,b) is the incident wave to the right, while (b,x) is taken to be the fundamental *reflected* wave to the left. Its coefficient is the *reflection coefficient* r, namely

$$r = i[a,b]^2 = i\exp\left(2ih\int_a^b q^{1\ 2}\,dz\right). \qquad (4.14)$$

If $a = 0$ and if q is real for all x, then

$$r = i[0,b]^2 = i\exp\left(-2ih\int_b^0 q^{1/2}\,dz\right), \qquad (4.15)$$

where $[0,b]$ is the exponential of a purely imaginary quantity. Hence $|r| = 1$.

The actual mathematical process of reflection takes place in the neighbourhood of the point $z = a$, where the functional character of the solution undergoes substantial metamorphosis. Physically, along the real axis, the process takes place in the somewhat vague range nearest to the point $z = a$. The fact that the phase-integral occurring in (4.15) is extended up to $z = 0$ has led many writers to say that the wave actually reaches $z = 0$, being reflected by some mirror-like agency, but this is a misinterpretation of equation (4.15). The upper limit $z = 0$ originates from a change of reference level, and in no way indicates that the wave reaches that point; in fact, the W.K.B.J. solutions cease to be valid near $z = 0$, and the more complicated solution (2.10) takes over there. In the process of reflection, the subdominant solution is merely moulded into progressive waves near the transition point.

4.7 Example
We consider the equation

$$w'' + h^2(1 - e^{i\epsilon} e^{kz}) w = 0$$

with $0 \leqslant \epsilon \leqslant \frac{1}{2}\pi$, $k > 0$. Transition points occur when

$$q \equiv 1 - e^{i\epsilon} e^{kz} = 0,$$

namely when $z = i(2\pi n - \epsilon)/k$, ($n$ integral). An infinite set of transition points exists along the imaginary axis, spaced by the interval $2\pi i/k$. Only one anti-Stokes line is asymptotic to $-Ox$, namely one from the transition point given by $n = 0$, as in Fig. 15. (Considerable inspection

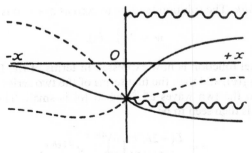

FIG. 15

of the W.K.B.J. solutions may be necessary to come to this conclusion).

With the branch cuts inserted as shown, we may use the solutions found in example 1, section 2.7, by replacing the z in that example by $z + i\epsilon/k$. Then

$$(-i\epsilon/k, x) = -i \exp [2ih(-1+\log 2)/k] e^{-ih(x+i\epsilon/k)}, \quad (\rightarrow)$$

$$(x, -i\epsilon/k) = -i \exp [-2ih(-1+\log 2)/k] e^{ih(x+i\epsilon/k)} \quad (\leftarrow)$$

along $-Ox$. Then $r = i$ from equation (4.14) with b taken at the transition point.

Using $e^{\pm ihx}$ as the fundamental waves, we modify r thus:

$$r = i \exp [-4ih(-1+\log 2)/k] e^{-2h\epsilon/k}. \qquad (4.16)$$

The particular equation we have chosen is one for which an exact solution may be obtained; briefly, we proceed as follows. If the independent variable z is changed to v, where

$$v = (2hi/k) e^{kz/2} e^{i\epsilon/2},$$

we obtain Bessel's equation

$$w'' + v^{-1} w' + [1 - (2hi/k)^2 v^{-2}] w = 0$$

of order $2hi/k$. The solution tending to zero as $z \rightarrow +\infty$ is

$$w = H^{(1)}_{2hi/k}(v).$$

This Hankel function is now expressed in terms of the Bessel functions $J_{\pm 2hi/k}(v)$, and then the first terms of the two series expansions give rise to the two waves $e^{\pm ihz}$ when $|v|$ is small. The reflection coefficient is then seen to be

$$r = \frac{\Gamma(-2hi/k)}{\Gamma(2hi/k)} \left(\frac{h}{k}\right)^{4hi/k} e^{-2\epsilon h/k}.$$

If h is large or k small, we may use the asymptotic expression for the gamma function, interpreting $\pm i = e^{\pm i\pi/2}$. Using

$$\Gamma(z) \sim \sqrt{(2\pi)}\, e^{-z} z^{z-(1/2)} \qquad |\arg z| < \pi$$

(see the Appendix), we obtain

$$r \doteqdot \frac{e^{2hi/k}(2he^{-i\pi/2}/k)^{-2hi/k-(1/2)}}{e^{-2hi/k}(2he^{i\pi/2}/k)^{2hi/k-(1/2)}}\left(\frac{h}{k}\right)^{4hi/k} e^{-2\epsilon h/k}$$

$$= i2^{-4hi/k}e^{4hi/k}e^{-2\epsilon h/k}$$

identical with formula (4.16) found by phase-integral methods.

Clearly $|r| = e^{-2h\epsilon/k} < 1$, unless $\epsilon = 0$, when the medium is loss-free.

4.8 Reflection coefficients by phase integration
Guided by remarkable successes in the older quantum theory, Eckersley has always used different considerations for obtaining the reflected wave, though providing no proof of the alternative method used. The justification of his method appears to be as follows.

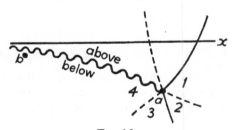

FIG. 16

We swing the branch cut round to coincide with the negative real axis, and consider the incident and reflected waves just below the cut. The subdominant solution $w = (z,b)_s$ in domains 1 and 2 becomes

1, 2: $(z,b)_s \equiv [a,b](z,a)_s$,
3:      $[a,b](z,a)_d$,
4:      $[a,b](z,a)_d - i[a,b](a,z)_s$
        $\equiv (z,b)_d - i[a,b]^2(b,z)_s$.

Since (b, z) is the incident wave, we divide by $-i[a, b]^2$, obtaining

$$w = (b, z) + i[b, a]^2(z, b)$$

$$w = q^{-1/4}\{[b, z] + i[b, a]^2 [z, b]\},$$

since in keeping with the notation introduced in section 2.7, we may write

$$(b, z) \equiv q^{-1/4}[b, z], \qquad (z, b) \equiv q^{-1/4}[z, b],$$

All the integrals are evaluated just below the branch cut. On the other hand, if $[z, b]$ and one of the $[b, a]$ are evaluated just above the cut (the factor $q^{-1/4}$ remaining the same), numerical identity is maintained by writing

$$\underset{\text{(below)}}{[z, b]} = \underset{\text{(above)}}{[b, z]}, \qquad \underset{\text{(below)}}{[b, a]} = \underset{\text{(above)}}{[a, b]},$$

since $q^{1/2}$ in the integrand merely changes sign. Hence

$$w = q^{-1/4}\{\underset{\text{(below)}}{[b, z]} + i\underset{\text{(below)}}{[b, a]}\,\underset{\text{(above)}}{[a, b]}\,\underset{\text{(above)}}{[b, z]}\}$$

$$= q^{-1/4}\{\underset{\text{(below)}}{[b, z]} + i\underset{\text{(anti-clockwise)}}{[b, z]}\}$$

since $\underset{\text{(above)}}{[b, a]}\,\underset{\text{(above)}}{[a, b]}\,\underset{\text{(above)}}{[b, z]}$ is merely the exponential of the integral taken right round from b (below the cut) to z (above the cut), as in Fig. 17. Cauchy's theorem then enables us to detach the integral $\underset{\text{(anti-clockwise)}}{[b, z]}$ from $z = a$, distorting it into a loop.

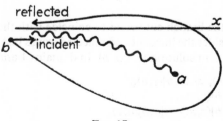

FIG. 17

In other words, the incident wave $q^{-1/4}[b, z]$ is transformed into the appropriate reflected wave (complete with reflection coefficient apart from the phase factor i) by integrating right round the loop, the factor $q^{-1/4}$ remaining the same. This loop integral, known as a *complex phase-integral,* provides the name to the subject.

4.9 Conservation of energy flow through a transition point
In case (ii), section 4.5, we have seen that the subdominant term on a Stokes line remains undefined in the presence of a dominant term. But the principle of conservation of energy flow in a loss-free medium enables us to be more definite regarding this subdominant term.

Using Fig. 13, we have for a general solution

$$w = A(0, x) + B(x, 0)$$

along the anti-Stokes line OB:

4: $A(0, z)_s + B(z, 0)_d$,
3: $(A - iB)(0, z)_s + B(z, 0)_d$,
2: $(A - iB)(0, z)_d + B(z, 0)_s$,
1: $(A - iB)(0, z)_d - iA(z, 0)_s$.

On the Stokes line OA, we shall postulate the addition (in the positive sense) of a multiple Ti times the dominant coefficient to the subdominant coefficient, where T is to be found if possible. Hence

$$OA: (A - iB)(0, x)_d + [B - iT(A - iB)](x, 0)_s.$$

We shall now use the conservation equation (4.7), namely

$$\mathrm{Im}\,(w'\,w^*) = \text{constant}.$$

Along OA, let $\arg q = -\pi$ and along OB, let $\arg q = 0$. Then along OB

$$w^* = A^*(x, 0) + B^*(0, x),$$

and $\qquad w' = Ai\,hq^{1/2}(0, x) - Bi\,hq^{1/2}(x, 0)$

to the order of differentiating the exponentials only. Hence

$$w'\,w^* = ihq^{1/2}[A(0, x) - B(x, 0)][A^*(x, 0) + B^*(0, x)]$$
$$= ih\{AA^* - BB^* + AB^*[0, x]^2 - BA^*[x, 0]^2\}.$$

Now $AB^*[0,x]^2$ and $BA^*[x,0]^2$ are conjugate quantities, so their difference is imaginary. Hence

$$\text{Im}\,(w'\,w^*) = h(AA^* - BB^*). \tag{4.17}$$

Along OA, where $\arg q = -\pi$, it can easily be checked that

$$(0,x)_d^* = -i(0,x)_d, \qquad (x,0)_s^* = -i(x,0)_s.$$

Hence

$$
\begin{aligned}
w'\,w^* &= ihq^{1/2}\{(A-iB)(0,x)_d - [B-iT(A-iB)](x,0)_s\} \times \\
&\quad \times (-i)\{(A^*+iB^*)(0,x)_d + [B^*+iT^*(A^*+iB^*)](x,0)_s\} \\
&= h\{(A-iB)[0,x]_d - [B-iT(A-iB)][x,0]_s\} \times \\
&\quad \times \{(A^*+iB^*)[0,x]_d + [B^*+iT^*(A^*+iB^*)][x,0]_s\} \\
&= h\{(A-iB)[B^*+iT^*(A^*+iB^*)] - (A^*+iB^*) \times \\
&\quad \times [B-iT(A-iB)] + \text{real quantities}\},
\end{aligned}
$$

from which it follows that

$$
\begin{aligned}
\text{Im}\,(w'\,w^*) &= 2h\,\text{Im}\,\{(A-iB)[B^*+iT^*(A^*+iB^*)]\} \\
&= 2h[\text{Im}\,(AB^*) - BB^* + (\text{Rl}\,T)(AA^*+BB^*) - \\
&\quad - (\text{Rl}\,T).2\,\text{Im}\,(AB^*)]. \tag{4.18}
\end{aligned}
$$

Equating (4.17) and (4.18), we obtain after rearrangement

$$[AA^* + BB^* - 2\,\text{Im}\,(AB^*)](1 - 2\,\text{Rl}\,T) = 0,$$

implying that $\text{Rl}\,T = \tfrac{1}{2}$, unless

$$AA^* + BB^* - 2\,\text{Im}\,(AB^*) = 0,$$

or

$$(\text{Rl}\,A - \text{Im}\,B)^2 + (\text{Im}\,A - \text{Rl}\,B)^2 = 0,$$

or

$$\text{Rl}\,A = \text{Im}\,B, \qquad \text{Im}\,A = \text{Rl}\,B,$$

or

$$B = iA^*,$$

which merely excludes the case in which there is no energy flow, namely when $AA^* - BB^* = 0$.

Hence any Stokes constant of the form $i(\frac{1}{2} + ik)$ where k is real and arbitrary preserves the flow of energy through the transition point. In particular, if $k = 0$, yielding the constant $\frac{1}{2}i$, we have the simplest Stokes constant that preserves energy flow. This does not of course define an actual value of the subdominant term on the Stokes line. It can be seen immediately that this particular value averages the values of the two subdominant coefficients on each side of the Stokes line.

Now the second version of Jeffreys' connection formula (4.9) has in effect implicitly used this value of the Stokes constant on the Stokes line, since we have already seen in section 4.5 case (ii) that the average value of the two subdominant terms on each side of the line is zero. Hence the use of Jeffreys' second formula in the wrong direction will preserve energy flow through the transition point.

This vital fact, which appears to have escaped the notice of writers on phase-integral methods, explains why Jeffreys should state that no wrong results have been reached from the use of the formula, since the problems considered have related to loss-free media (see Chapter V for further details). The present author has always maintained that the first use of formula (4.9) backwards (B. Jeffreys [63]) yielded the correct answer by a happy coincidence.

It should be pointed out that even in the case of no energy flow we may still use $T = \frac{1}{2}$, though this is not necessary.

4.10 Transition points of order n

If q possesses a zero of multiplicity n, the above theory using a Stokes constant i is not valid. This case has been considered by Goldstein [46], but we consider the equation here using our present apparatus.

The comparison equation used in section 2.2 must now be taken to be

$$d^2 X/d\xi^2 = \xi^n X, \tag{4.19}$$

and the asymptotic expressions of this equation will yield the usual W.K.B.J. solutions of the original equation. The Stokes constants for this equation (4.19) will therefore be the Stokes constants for the original equation.

If we now let

$$X = \xi^{1/2} Y, \qquad \eta = 2i\xi^{(n+2)/2}/(n+2),$$

equation (4.19) transforms easily into Bessel's equation of order $1/(n+2)$:

$$\frac{d^2 Y}{d\eta^2} + \frac{1}{\eta} \cdot \frac{dY}{d\eta} + \left(1 - \frac{[1/(n+2)]^2}{\eta^2}\right) Y = 0,$$

with a solution

$$X = \xi^{1/2} J_{1/(n+2)}[2i\xi^{(n+2)/2}/(n+2)]$$

whose Stokes constants are $2i\cos[\pi/(n+2)]$ according to equation (3.14).

Hence $T = 2i\cos[\pi/(n+2)]$ for all Stokes lines radiating outwards from the transition point of order n of the original equation. In particular, if the order is 2, we have

$$T = 2i\cos\tfrac{1}{4}\pi = i\sqrt{2}. \tag{4.20}$$

Anti-Stokes lines occur when

$$\mathrm{Rl}\, ih \int_0^z q^{1/2}\, dz = 0.$$

If $q = z^n$ (perhaps with a constant factor), the lines locally are given by

$$\mathrm{Rl}\, iz^{(n+2)/2} = 0,$$

yielding $(n+2)$ such lines evenly spaced around the transition point. If $n = 2$, there are four anti-Stokes lines.

Rules for crossing a branch cut. If $w = (0,z)$ occurs just before a branch cut where $z = re^{i\delta}$, then

$$q = (re^{i\delta})^n f(z) \quad \text{and} \quad q^{1/2} = (re^{i\delta})^{n/2} f^{1/2}$$

where $f^{1/2}$ is single-valued around $z = 0$. Then

$$(0,z) = (re^{i\delta})^{-n/4} f^{-1/4} \exp\left(ih \int_0^z (re^{i\delta})^{1/2} f^{1/2}\, dz\right).$$

Just after the branch cut, $z = e^{i\delta - 2\pi i}$, so this solution must be replaced by

$$(ze^{2\pi i})^{-n/4} f^{-1/4} \exp\left(ih \int_0^z (ze^{2\pi i})^{1/2} f^{1/2} \, dz \right)$$

$$\left. \begin{aligned} &= e^{-n\pi i/2}(z, 0) \equiv i(-1)^{(n+1)/2}(z, 0) && \text{if } n \text{ is odd,} \\ &\text{or } e^{-n\pi i/2}(0, z) \equiv (-1)^{n/2}(0, z) && \text{if } n \text{ is even,} \end{aligned} \right\} \quad (4.21)$$

with similar rules for $(z, 0)$.

In particular, consider $q(x)$ real for real x with a transition point of order 2 at $x = 0$.

Case (i). Anti-Stokes lines along $\pm Ox$.

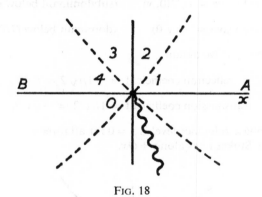

Fig. 18

Along OA, we choose the general solution

$$w = C(0, x) + D(x, 0)$$

with the usual error terms implied. Then

$$
\begin{aligned}
1: \quad & w = C(0, z)_s + D(z, 0)_d, \\
2: \quad & (C + iD\sqrt{2})(0, z)_s + D(z, 0)_d, \\
3: \quad & (C + iD\sqrt{2})(0, z)_d + D(z, 0)_s, \\
4: \quad & (C + iD\sqrt{2})(0, z)_d + [D + i\sqrt{2}(C + iD\sqrt{2})](z, 0)_s, \\
OB: \quad & (C + iD\sqrt{2})(0, x) + (iC\sqrt{2} - D)(x, 0),
\end{aligned}
$$

yielding the two connection formulae:

$$(0, x) + i\sqrt{2}(x, 0) \leftrightarrow (0, x),$$

$$i\sqrt{2}(0, x) - (x, 0) \leftrightarrow (x, 0),$$

and also

$$(0, x) \leftrightarrow -(0, x) - i\sqrt{2}(x, 0),$$

$$(x, 0) \leftrightarrow (x, 0) - i\sqrt{2}(0, x).$$

For a transmitted wave to the right along $+Ox$, $C = 0$, since $(0, x)$ is the wave transmitted to the left (dominant below OA). Hence, using the second connection formula, along OB we have

incident wave $= i\sqrt{2}(0, x)$ (subdominant below OB),

reflected wave $= -(x, 0)$ (dominant below OB).

Dividing by $i\sqrt{2}$, we obtain

reflection coefficient $= -1/i\sqrt{2} = i/\sqrt{2}$,

transmission coefficient $= 1/i\sqrt{2} = -i/\sqrt{2}$,

when the phase-reference level is $x = 0$ for all three waves.

Case (ii). Stokes lines along $\pm Ox$.

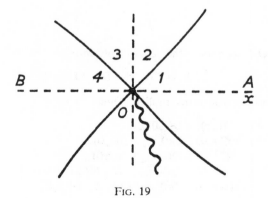

Fig. 19

Along OA, let $(x,0)_s$ be a subdominant solution. Then

1: $(z,0)_s$,
2: $(z,0)_d$,
3: $(z,0)_d + i\sqrt{2}(0,z)_s$,
4: $(z,0)_s + i\sqrt{2}(0,z)_d$.

Hence we have the connection formula

$$i\sqrt{2}(0,x)_d \leftarrow (x,0)_s.$$

Further connection formulae may easily be manufactured.

CHAPTER V

Two Transition Points

5.1 Introduction

The discussion of the equation

$$w'' + h^2 q(z, h) w = 0 \tag{5.1}$$

through a domain in which q has two complex zeros at $z = \pm a$ is more difficult. If $q(z, h)$ has two zeros only, it is merely proportional to $z^2 - a^2$, for which case an exact solution is obtainable. But if q is of the form

$$q = (z^2 - a^2) r(z),$$

where $r(z)$ contains further zeros, phase-integral methods are of assistance in the solution of equation (5.1).

We shall usually assume that $r(z)$ is such that q tends to a constant limit as $z \to \pm \infty$ along the real axis; hence anti-Stokes lines exist along these axes at infinity, and we assume that these lines originate from one or other or both of the transition points $z = \pm a$. We also postulate that the zeros of r are such that the metamorphosis of the solution w as we pass from $x = -\infty$ to $x = +\infty$ along the real axis is determined in the W.K.B.J. sense solely by the two zeros $z = \pm a$.

Various cases arise:

(i) $q(x)$ is real for real x, with two real zeros at $x = \pm a$ such that the medium is overdense ($q < 0$) between the zeros; r is real and positive along the real axis. This is the problem of the *overdense potential barrier*.

(ii) $q(x)$ is real and positive for all real x, with two complex zeros at $z = \pm ib$ (b real), apart from further zeros of $r(z)$, which again is real and positive on the real axis. This is the problem of the *underdense potential barrier*.

(iii) $q(x)$ tends to a real positive constant as $x \to \pm \infty$, but is complex elsewhere along the real axis, possessing two transition points at the complex points $z = \pm a$.

(iv) $q(x)$ is real for real x, tending to $-\infty$ as $x \to \pm \infty$, but positive between the two real transition points $x = \pm a$. This is the problem of the approximate *harmonic oscillator*, for which the *eigenvalues* of h are required.

(v) More general profiles of $q(x)$ with transition points of order greater than unity.

5.2 The overdense potential barrier

This case has been described by B. Jeffreys [63] and later by Jeffreys [67] and other workers. The profile of q (real) along the real axis is shown in Fig. 20. Evidently evanescent W.K.B.J. solutions exist for

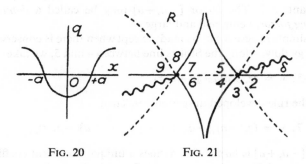

FIG. 20   FIG. 21

$-a < x < a$. In the complex z-plane, we construct Stokes and anti-Stokes lines emerging from both $z = -a$ and $+a$; evidently a Stokes line exists along the real axis joining the two transition points. The lines cannot necessarily be extended further than indicated owing to the presence of other zeros but we assume that q is such that solutions uniformly asymptotic in the sense of Chapter II exist in the domain shown. Branch cuts are inserted up to $z = a$ at a small positive angle δ to $+Ox$, and up to $z = -a$ at the small angle δ to $-Ox$ for symmetrical convenience, though strictly speaking these cuts are quite arbitrary in position.

Along the $+Ox$ axis a wave to the right is represented by $(x, +a)$

with respect to the phase reference level $x = +a$; this becomes subdominant in domain 2. Tracing the solution round, we obtain

2: $(z, +a)_s$,
3: $(z, +a)_s$,
4: $(z, +a)_d$,

No unique subdominant solution is defined in domains 4 and 5, but an attempt at its specification is essential. When transferred to the reference level $z = -a$, $(z, +a)_d$ becomes $(z, -a)_s[-a, +a]$, where $(z, -a)_s$ is now subdominant in 6 and 7 with reference to $(-a, z)_d$. That this change takes place may be seen from the fact that $(z, +a)_d$ is dominant in 5 and in its extension out to the domain marked R. Hence $(z, -a)_d[-a, +a]$ is also dominant in R, so $(z, -a)_s$ is subdominant in 7. The factor $[-a, +a]$ may be called a *dominancy changing factor*, being real and large.

No simple progress can be made except when there is conservation of energy flow. Along the Stokes line between 4 and 5, we take

4–5: $w = (z, +a)_d - \frac{1}{2}i(+a, z)_s$,

using the rule developed in section 4.9. Then

6, 7: $w = (z, -a)_s[-a, +a] - \frac{1}{2}i[+a, -a](-a, z)_d$,

where $[-a, +a]$ is large. This defines a unique dominant coefficient $-\frac{1}{2}i[+a, -a]$. The subdominant coefficient in 7 must be

$$[-a, +a] + \frac{1}{2}i\{-\frac{1}{2}i[+a, -a]\},$$

where a Stokes constant $\frac{1}{2}i$ is used in passing from a Stokes line to the neighbouring domain. Hence

7: $\{[-a, +a] + \frac{1}{4}[+a, -a]\}(z, -a)_s - \frac{1}{2}i[+a, -a](-a, z)_d$,
8: $\{[-a, +a] + \frac{1}{4}[+a, -a]\}(z, -a)_d - \frac{1}{2}i[+a, -a](-a, z)_s$,
9: $\{[-a, +a] + \frac{1}{4}[+a, -a]\}(z, -a)_d +$
$\qquad + i\{[-a, +a] - \frac{1}{4}[+a, -a]\}(-a, z)_s$.

The incident wave is therefore

$$\{[-a, +a] + \frac{1}{4}[+a, -a]\}(x, -a)$$

and the reflected wave

$$i\{[-a, +a] - \tfrac{1}{4}[+a, -a]\}(-a, x)$$

referred to the phase-reference level $x = -a$. Hence the reflection coefficient r is given by

$$r = i\frac{[-a, +a] - \tfrac{1}{4}[+a, -a]}{[-a, +a] + \tfrac{1}{4}[+a, -a]} = i\{1 - \tfrac{1}{2}[+a, -a]^2\} \qquad (5.2)$$

since $[+a, -a]$ is real and small, and the transmission coefficient t is given by

$$t = \frac{1}{[-a, +a] + \tfrac{1}{4}[+a, -a]} = [+a, -a], \qquad (5.3)$$

both to the second order in $[+a, -a]$. This method is equivalent to the use of Jeffreys' second connection formula (4.9); the use of this formula in the normally forbidden sense is permitted only on account of the conservation of energy flow. Jeffreys [67] does not seem to have appreciated this point even in his later papers.

The moduli $|r|$ and $|t|$ may be obtained more accurately, in such a way that the result is independent of whether or not the W.K.B.J. solutions are valid between the two transition points. The two transition points $\pm a$ form a group, and it must be assumed that further zeros of q do not prevent the anti-Stokes lines grouping themselves as shown in Fig. 22. If this is not so, the following theory is of course

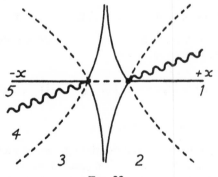

Fig. 22

invalidated. More specifically, we assume that uniformly asymptotic solutions of equation (5.1) exist throughout the domain of Fig. 22 (including the points $\pm a$) expressed in terms of the Weber parabolic cylinder functions.

For a wave transmitted to the right along the positive real axis, we have $w = (z, +a)_s$ in domain 1. The assumptions just stated enable us to trace this solution round a large semicircle below the transition points, using T as the unknown Stokes constant across the Stokes line between domains 3 and 4. We have

1 : $w = (z, +a)_s$,
2 : $(z, +a)_s$,
3 : $(z, +a)_d \equiv [-a, +a](z, -a)_d$,
4 : $[-a, +a](z, -a)_d - T[-a, +a](-a, z)_s$,
5 : $i[-a, +a](-a, z)_d - iT[-a, +a](-a, z)_s$,

The latter is the incident wave, so

$$r = 1/T, \qquad t = i/T[-a, +a].$$

Since $[-a, +a]$ is real along the real axis, we obtain

$$|r|^2 = 1/TT^*, \qquad |t|^2 = [+a, -a]^2/TT^*. \qquad (5.4)$$

To find TT^*, we use equation (4.7) to connect up the solutions along the positive and negative real axes, namely

$$w = (x, +a) \qquad\qquad x \gg a,$$

$$w = i[-a, +a]\{(-a, x) - T(x, -a)\} \qquad x \ll -a.$$

With the branch cuts as shown, we may choose $\arg q = 0$ for $x > a$ and for $x < -a$; hence for $x \gg a$,

$$(x, +a)^* = \left[q^{-1/4} \exp\left(ih \int_x^a q^{1/2}\, dx \right) \right]^* = (+a, x),$$

and $\qquad (x, +a)' = -ihq^{1/2}(x, +a)$,

to the accuracy inherent in differentiating the exponential only. The term containing the derivative q' is neglected, but if as $x \to \infty$, q tends to a constant, no approximation is in fact being made.

For $x \ll -a$, similar formulae apply, together with

$$(-a, x)^* = (x, -a), \qquad (-a, x)' = ihq^{1/2}(-a, x).$$

Hence equation (4.7) yields, (with $[-a, +a]$ real),

$$\text{Im } i[-a, +a]\{ihq^{1/2}(-a, x) + Tihq^{1/2}(x, -a)\}(-i)[-a, +a] \times$$
$$\times \{(x, -a) - T^*(-a, x)\} = \text{Im } \{-ihq^{1/2}(x, +a)(+a, x)\}.$$

Now $(x, \pm a)(\pm a, x)q^{1/2} = 1$, so

$$[-a, +a]^2 \text{Im } \{i - iTT^* - iT^*q^{1/2}(-a, x)^2 +$$
$$+ iTq^{1/2}(x, -a)^2\} = \text{Im } (-i);$$

but $-iT^*q^{1/2}(-a, x)^2$ and $iTq^{1/2}(x, -a)^2$ are conjugate quantities, so

$$[-a, +a]^2(1 - TT^*) = -1,$$

yielding
$$TT^* = 1 + [+a, -a]^2.$$

Here, $[+a, -a]^2$ is usually small, unless the two zeros are close together. Hence results (5.4) yield

$$|r| = 1/\sqrt{\{1 + [+a, -a]^2\}};$$
$$|t| = [+a, -a]/\sqrt{\{1 + [+a, -a]^2\}}. \tag{5.5}$$

The values (5.2) and (5.3) already found are merely the expansions of these expressions to the second order in $[+a, -a]$. As far as r itself is concerned, the phase factor i occurring in (5.2) is that which would be obtained had the lower transition point $z = -a$ alone been used, employing the technique of section 4.6.

In formulae (5.5), a may tend to zero, yielding $|r| = |t| = 1/\sqrt{2}$, as found in section 4.10.

5.3 The underdense potential barrier

The barrier for which $q = (x^2 + b^2) r(x) > 0$ for all real x contains two important transition points $z = \pm ib$ in the complex plane; this plane appears as depicted in Fig. 23.

If the upper zero $z = +ib$ is considered first, a uniform asymptotic solution similar to (2.10) exists in domains 5, 6, 7, 8, for which

Stokes constants i may be used. The wave transmitted to the right along $+Ox$ may be taken to be $(x, 0)$, taking O as the phase reference level. Then $(z, 0)_d$ is dominant in 5; hence

5: $w = [ib, 0](z, ib)_d,$
6: $\qquad [ib, 0](z, ib)_s,$
7: $\qquad [ib, 0](z, ib)_s,$
8: $\qquad [ib, 0](z, ib)_d \equiv (z, 0)_d,$

representing an incident wave only. Hence $r = 0$, $t = 1$, denoting that the barrier is completely transparent. Ordinary ray theory along the

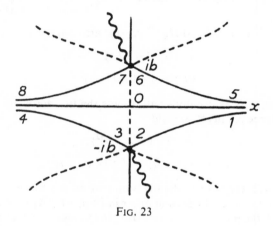

Fig. 23

real axis would suggest this result, since, in optical terms, the refractive index nowhere vanishes along the path.

An increase of accuracy can be obtained if the lower transition point $z = -ib$ is used. Since a uniform asymptotic solution of the form (2.10) exists around $z = -ib$ in domains 1, 2, 3, 4, we have

1: $w = [-ib, 0](z, -ib)_s,$
2: $\qquad [-ib, 0](z, -ib)_d,$
3: $\qquad [-ib, 0](z, -ib)_d + i[-ib, 0](-ib, z)_s,$
4: $\qquad [-ib, 0](z, -ib)_s + i[-ib, 0](-ib, z)_d,$
$\qquad \equiv (z, 0)_s + i[-ib, 0]^2(0, z)_d,$

the former being the incident wave. Hence

$$r = i[-ib, 0]^2, \qquad t = 1.$$

Usually, $[-ib, 0]$ is approximately equal to $[0, ib]$, so $r = i[-ib, ib]$ where $[-ib, ib]$ is real and small.

If both transition points are used, a method may be designed similar to that given in the previous section, whereby further accuracy is obtained. The results are:

$$r = i[-ib, ib], \qquad t = 1 - \tfrac{1}{2}[-ib, ib]^2.$$

Moreover, since there is conservation of energy flow, more exact values of $|r|$ and $|t|$ may be found as in the previous section; these are

$$|r| = [-ib, ib]/\sqrt{\{1 + [-ib, ib]^2\}},$$
$$|t| = 1/\sqrt{\{1 + [-ib, ib]^2\}},$$

valid also when b is small; these demonstrate the nature of the approximations just obtained.

5.4 The complex barrier

When $q = (z^2 - a^2)r(z)$, in which a is complex with $0 \leqslant \arg a \leqslant \tfrac{1}{2}\pi$, the methods given in sections 5.2 and 5.3 are no longer applicable. Only the main trend of the method by which the problem may be solved is presented here. We assume that there is a domain similar to that shown in Fig. 22 (but with the transition points in complex positions), throughout which uniform asymptotic solutions exist that may be expressed in terms of the Weber parabolic cylinder functions.

Anti-Stokes lines exist along lines defined by $\arg z = 0, \tfrac{1}{2}\pi, \pi, \tfrac{3}{2}\pi$ for $|z| \gg |a|$, and the Stokes constants discussed in section 3.7 may be used. Modifications, however, are necessary, since it will be recalled that a vague phase-reference level was employed in that section. Here, we must be more precise, and this introduces several complicated numerical factors, which are not derived here.

If branch cuts are inserted from $z = a$ along $\arg(z - a) = \delta$ and from $z = -a$ along $\arg(z + a) = \pi + \delta$ (δ small and positive), and if $\arg q = 0$

for large real positive x, then it may be demonstrated that the appropriate connection formulae are

$$(x, -a) + r(-a, x) \leftrightarrow t(x, +a), \tag{5.7}$$

$$t(-a, x) \leftrightarrow (+a, x) + r(x, +a), \tag{5.8}$$

where waves along $-Ox$ are measured with respect to the complex phase-reference height $z = -a$ and waves along $+Ox$ with respect to $z = +a$. Here, we have

$$r = i(2\pi)^{-1/2} \, \Gamma(\tfrac{1}{2} - \tfrac{1}{2}ia^2 h) \exp\left(-\tfrac{1}{2}ia^2 h\right) \exp\left(\tfrac{1}{4}\pi a^2 h\right)(\tfrac{1}{2}a^2 h)^{ia^2h/2}, \tag{5.9}$$

$$t = -ir \exp\left(-\tfrac{1}{2}\pi a^2 h\right). \tag{5.10}$$

The mathematician may easily check that the first term of the asymptotic development of r when $|a^2 h|$ is large is merely $r \sim i$. This is the result given by phase-integral methods using the transition point $z = -a$ only; result (5.9) is valid even when $|a^2 h|$ is small. In particular, when $a = 0$, $r = i(2\pi)^{-1/2} \, \Gamma(\tfrac{1}{2}) = i/\sqrt{2}$ and $t = 1/\sqrt{2}$, results similar to those found in section 4.10, but bearing in mind that the branch cuts are in different positions.

If $\arg a = 0$, we may calculate the modulus of r; this equals

$$|r| = (2\pi)^{-1/2} \, |\Gamma(\tfrac{1}{2} - \tfrac{1}{2}ia^2 h)| \exp\left(\tfrac{1}{4}\pi a^2 h\right),$$

or
$$|r|^2 = (2\pi)^{-1} \, \Gamma(\tfrac{1}{2} - \tfrac{1}{2}ia^2 h) \, \Gamma(\tfrac{1}{2} + \tfrac{1}{2}ia^2 h) \exp\left(\tfrac{1}{2}\pi a^2 h\right)$$

$$= (2\pi)^{-1} \, \pi \exp\left(\tfrac{1}{2}\pi a^2 h\right)/\sin\left[\pi(\tfrac{1}{2} - \tfrac{1}{2}ia^2 h)\right]$$

$$= 1/[1 + \exp\left(-\pi a^2 h\right)],$$

in keeping with formula (5.5).

5.5 The approximate harmonic oscillator
Let $q(x)$ be real for all x, and underdense ($q > 0$) between the two real transition points $x = \pm a$. If further zeros of $q(z)$ are well displaced from the points $\pm a$, the disposition of lines in the complex plane appears as shown in Fig. 24, in which an anti-Stokes line joins the two zeros. The eigenvalues of h are required, such that subdominant

solutions exist for large $|x|$ along both the positive and negative real axes. Conversely, for given large h, we seek the eigenvalues of a.

If it is assumed that the W.K.B.J. solutions are valid between the zeros, we proceed as follows: The subdominant solution in domain 1 of the differential equation

$$w'' + h^2(a^2 - z^2)\, r(x)\, w = 0$$

is $w = (+a, x)_s$, if $\arg q = 0$ for $-a < x < +a$, and $\arg q = \pi$ for

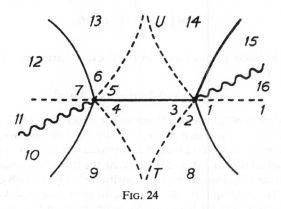

FIG. 24

$x > +a$. Tracing the solution round between the zeros, we obtain

1: $(+a, z)_s$,   (5.11)

2: $(+a, z)_d$,

3: $(+a, z)_d - i(z, +a)_s$,   (5.12)

5: $[+a, -a](-a, z)_s - i[-a, +a](z, -a)_d$,

6: $\{[+a, -a] + [-a, +a]\}(-a, z)_s - i[-a, +a](z, -a)_d$,

7: $\{[+a, -a] + [-a, +a]\}(-a, z)_d - i[-a, +a](z, -a)_s$.

This solution is subdominant along the negative real axis if

$$[+a, -a] + [-a, +a] = 0,$$

or $$[-a, +a]^2 = -1.$$ (5.13)

In terms of q, this condition yields

$$\exp\left(2ih \int\limits_{-a}^{+a} q^{1/2} \, dx\right) = -1,$$

or

$$2ih \int\limits_{-a}^{+a} q^{1/2} \, dx = i(\pi + 2n\pi),$$

so

$$h = \pi(\tfrac{1}{2} + n) \left/ \int\limits_{-a}^{+a} a^{1/2} \, dx \right.,$$

where n is integral and large. For the exact harmonic oscillator, $^2 = (a^2 - x^2)$, so

$$h = \pi(\tfrac{1}{2} + n)/\tfrac{1}{2}\pi a^2 = (1 + 2n)/a^2,$$

representing in effect the condition under which the solution of the equation may be expressed in terms of the *Hermite polynomials*.

On the other hand, if uniformly asymptotic solutions exist over the whole domain shown in Fig. 24 (including the transition points), then we may trace solutions round the transition points through domains 8, 9, ..., 16, yielding a result that is valid even when the W.K.B.J. solutions are not valid between the zeros. If T and U denote the Stokes constants (referred to $+a$, say) for the Stokes lines 8, 9 and 13, 14 respectively, we have

$$
\begin{aligned}
1: \quad & w = (+a, z)_s, \\
8: \quad & (+a, z)_d, \\
9: \quad & (+a, z)_d - T(z, +a)_s, \\
\equiv \quad & [+a, -a](-a, z)_d - T[-a, +a](z, -a)_s, \\
10: \quad & [+a, -a](-a, z)_s - T[-a, +a](z, -a)_d, \\
11: \quad & i[+a, -a](z, -a)_s - iT[-a, +a](-a, z)_d; \\
\text{and } 16: \quad & (+a, z)_s, \\
15: \quad & -i(z, +a)_s, \\
14: \quad & -i(z, +a)_d, \\
13: \quad & -i(z, +a)_d - iU(+a, z)_s, \\
12: \quad & -i[-a, +a](z, -a)_s - iU[+a, -a](-a, z)_d.
\end{aligned}
$$

We require equal subdominant solutions in domains 11 and 12 with no dominant terms; hence

$$T = U = 0, \qquad [+a, -a] = -[-a, +a],$$

the last equation being identical with (5.13), but now derived under more general conditions, namely that the two transition points may be close together if desired. This means that n may be any integer.

5.6 Normalization processes

When h has one of the values found in the last section it is usual to choose w such that

$$\int_{-\infty}^{+\infty} rww^* \, dx = 1.$$

To this end, we must evaluate this integral, in order to calculate the constant (apart from a phase factor) with which we must multiply w.

The elementary calculation (following Pauli [91], for example) is very unsatisfactory. Contributions to the integral from $x < -a$ and $x > +a$ are neglected, the solution being exponentially small. For $-a < x < +a$, the square of sines occurring in the integrand is replaced by $\frac{1}{2}$, that is by its mean on account of numerous oscillations for n large; moreover, with this approximation the W.K.B.J. solution is extended right up to the transition points, since the integral converges at these points.

According to (5.12), we have

$$w = (+a, x) - i(x, +a),$$

where arg $q = 0$. Hence

$$w^* = (x, +a) + i(+a, x),$$

and

$$\begin{aligned}
ww^* &= [(+a, x) - i(x, +a)][(x, +a) + i(+a, x)] \\
&= 2q^{-1/2} + i[(+a, x)^2 - (x, +a)^2] \\
&= 2q^{-1/2} + iq^{-1/2}\left[\exp\left(2ih\int_a^x q^{1/2}\,dx\right) - \exp\left(-2ih\int_a^x q^{1/2}\,dx\right)\right]
\end{aligned}$$

$$ww^* = 2q^{-1/2} - 2q^{-1/2}\sin\left(2h\int_a^x q^{1/2}\,dx\right)$$

$$= 2q^{-1/2}\left[1 - \cos\left(\tfrac{1}{2}\pi - 2h\int_a^x q^{1/2}\,dx\right)\right]$$

$$= 4q^{-1/2}\sin^2\left(\tfrac{1}{4}\pi - h\int_a^x q^{1/2}\,dx\right).$$

Hence

$$\int_{-\infty}^{+\infty} rww^*\,dx \doteqdot \int_{-a}^{+a} 4rq^{-1/2}\sin^2\left(\tfrac{1}{4}\pi - h\int_a^x q^{1/2}\,dx\right)dx$$

$$\doteqdot \int_{-a}^{+a} 2rq^{-1/2}\,dx, \tag{5.14}$$

using the average of the square of the sine. For the harmonic oscillator, we obtain

$$\int_{-\infty}^{+\infty} ww^*\,dx \doteqdot \int_{-a}^{+a}\frac{2\,dx}{\sqrt{(a^2-x^2)}} = 2\pi.$$

The following calculation, due to Furry [37], does not suffer from the defects of the previous integration.

Let w be that solution corresponding to the eigenvalue h. We now form two new solutions for slightly different values of a, but using the same value of h. Let w_1 be the solution if a_1 is used instead of a, such that w_1 is subdominant along the negative x-axis only and tending to w as $a_1 \to a$. Let w_2 be the solution if a_2 is used instead of a, such that w_2 is subdominant along the positive x-axis only and tending to w as $a_2 \to a$.

The differential equation yields

$$w'' w_1^* + h^2(a^2 - x^2)\,r(x)\,ww_1^* = 0,$$

and

$$w_1^{*\prime\prime} w + h^2(a_1^2 - x^2)\,r(x)\,w_1^* w = 0.$$

Subtraction yields

$$w'' w_1^* - w_1^{*''} w + h^2(a^2 - a_1^2) r(x) w w_1^* = 0,$$

or

$$\frac{d}{dx}(w' w_1^* - w_1^{*'} w) = h^2(a_1^2 - a^2) r(x) w w_1^*,$$

yielding upon integration

$$\left[w' w_1^* - w_1^{*'} w \right]_{-\infty}^{X} = h^2(a_1^2 - a^2) \int_{-\infty}^{X} r w w_1^* \, dx$$

where $-a < X < a$.

At the lower limit, both w and w_1^* vanish, so

$$w'(X) w_1^*(X) - w_1^{*'}(X) w(X) = h^2(a_1^2 - a^2) \int_{-\infty}^{X} r w w_1^* \, dx.$$

We now differentiate with respect to a_1, to obtain

$$w'(X) \frac{\partial}{\partial a_1} w_1^*(X) - w(X) \frac{\partial}{\partial a_1} w_1^{*'}(X) = 2h^2 a_1 \int_{-\infty}^{X} r w w_1^* \, dx +$$

$$+ \text{another term},$$

yielding when $a_1 = a$:

$$w'(X) \frac{\partial}{\partial a} w_1^*(X) - w(X) \frac{\partial}{\partial a} w_1^{*'}(X) = 2h^2 a \int_{-\infty}^{X} r w w^* \, dx.$$

Similarly,

$$w'(X) \frac{\partial}{\partial a} w_2^*(X) - w(X) \frac{\partial}{\partial a} w_2^{*'}(X) = -2h^2 a \int_{X}^{\infty} r w w^* \, dx,$$

so

$$\int_{-\infty}^{\infty} r w w^* \, dx = \frac{1}{2h^2 a} \left\{ w'(X) \frac{\partial}{\partial a} [w_1^*(X) - w_2^*(X)] - \right.$$

$$\left. - w(X) \frac{\partial}{\partial a} [w_1^{*'}(X) - w_2^{*'}(X)] \right\}. \tag{5.15}$$

Interchanging the order of differentiation in the last bracket, we may write this in the form

$$\left\{\frac{\partial}{\partial x}\left[\frac{\partial}{\partial a_1}w_1^*(x)-\frac{\partial}{\partial a_2}w_2^*(x)\right]_{a_1=a_2=a}\right\}_{x=X}.$$

From equation (5.12) we now take

$$w = (a,x)-i(x,a)$$

and

$$w_2 = (a_2,x)-i(x,a_2)$$

between the transition points, the value of w_2 for $x \gg a_2$ tending to $(a,x) \equiv w$ as $a_2 \to a$.

For w_1, we use in domain 7

7:   $w_1 = -i[-a_1,a_1](z,-a_1)_s,$
6:        $-i[-a_1,a_1](z,-a_1)_d,$
5:        $-[-a_1,a_1](-a_1,z)_s-i[-a_1,a_1](z,-a_1)_d,$

so for $-a_2 < x < a_2$ we take

$$w_1 = -[-a_1,a_1]^2(a_1,x)-i(x,a_1).$$

Hence

$$w_1-w_2 = -[-a_1,a_1]^2(a_1,x)-(a_2,x)-i(x,a_1)+i(x,a_2),$$

$$w_1^*-w_2^* = -[a_1,-a_1]^2(x,a_1)-(x,a_2)+i(a_1,x)-i(a_2,x).$$

Now $[\partial w_1^*/\partial a_1 - \partial w_2^*/\partial a_2]_{a_1=a_2=a}$ contains many terms that cancel, bearing in mind that $[a,-a]^2 = -1$. The only term that remains is

$$-(x,a)\left[\frac{\partial}{\partial a_1}[a_1,-a_1]^2\right] = -2(x,a)[a,-a]\frac{\partial}{\partial a}[a,-a].$$

But

$$\frac{\partial}{\partial a}[a,-a] = \frac{\partial}{\partial a}\exp\left(-ih\int_{-a}^{+a}q^{1/2}dx\right)$$

$$= [a,-a]\left(-ih\int_{-a}^{+a}\frac{\partial q^{1/2}}{\partial a}dx\right)$$

since q vanishes at both limits. Hence

$$\frac{\partial}{\partial a} w_1^* - \frac{\partial}{\partial a} w_2^* = -2ih(x,a) \int\limits_{-a}^{+a} \frac{\partial q^{1/2}}{\partial a} dx.$$

Equation (5.15) finally yields

$$\int\limits_{-\infty}^{+\infty} rww^* dx = \frac{1}{2h^2 a}\left[\frac{d}{dx}[(a,x)-i(x,a)] \times \left(-2ih(x,a) \int\limits_{-a}^{+a} \frac{\partial q^{1/2}}{\partial a} dx \right) - \right.$$

$$\left. - [(a,x)-i(x,a)] \times \frac{d}{dx}\left(-2ih(x,a) \int\limits_{-a}^{+a} \frac{\partial q^{1/2}}{\partial a} dx \right) \right]_{x=X}$$

$$= \frac{i}{ah} \int\limits_{-a}^{+a} \frac{\partial q^{1/2}}{\partial a} dx \left[-(a,x)'(x,a) + (a,x)(x,a)' \right]_{x=X}$$

$$= \frac{i}{ah} \int\limits_{-a}^{+a} \frac{\partial q^{1/2}}{\partial a} dx \left[-(ihq^{1/2} - \tfrac{1}{4}q'/q)q^{-1/2} + \right.$$

$$\left. + (-ihq^{1/2} - \tfrac{1}{4}q'/q)q^{-1/2} \right]_{x=X}$$

$$= \frac{2}{a} \int\limits_{-a}^{+a} \frac{\partial q^{1/2}}{\partial a} dx = \frac{2}{a} \int\limits_{-a}^{+a} r^{1/2} \frac{\partial}{\partial a}(a^2-x^2)^{1/2} dx$$

$$= 2 \int\limits_{-a}^{+a} rq^{-1/2} dx$$

as before (equation 5.14).

If $|a^2 h|$ is small, the above reasoning is not valid, since the W.K.B.J. solutions cannot be used between the transition points; a more comprehensive formula based on the Hermite polynomials may be obtained, but space forbids its inclusion here.

Sometimes, we may choose $q = f(x) - \lambda$ instead of $(x^2 - a^2) r(x)$, where q still possesses two zeros at $x = \pm a$. Under these circumstances, no weighting factor r enters the integrand of the normalised integral, and a similar calculation again yields the above result with the factor r omitted.

5.7 Transition points of order higher than unity

Example 1. Let $q(x)$ possess a double zero at $x = a$ and a single zero for $x = b > a$, such that $q(x) > 0$ for $x < a$ and for $a < x < b$, as in Fig. 25. Waves are transmitted through the medium in the underdense ranges $x < a$ and $a < x < b$, but the medium is overdense for $x > b$. We require to find the reflection coefficient in the range $x < a$ when the solution for $x > b$ is subdominant.

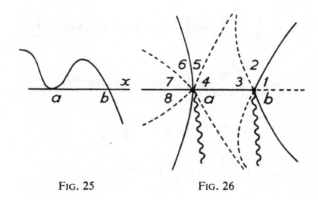

FIG. 25   FIG. 26

Further zeros of q are assumed not to affect the disposition of lines shown in Fig. 26. In domain 1, we have $w = (z, b)_s$. Hence, upon tracing the solution round (using Stokes constants i and $i\sqrt{2}$ for the two transition points respectively), we obtain

2: $(z, b)_d,$
3: $(z, b)_d + i(b, z)_s,$
4: $(z, a)_d [a, b] + i[b, a](a, z)_s,$

where $[a,b]$ is the exponential of a pure imaginary quantity,

5: $[a,b](z,a)_d + \{i[b,a]+i\sqrt{2}[a,b]\}(a,z)_s$,
6: $[a,b](z,a)_s + \{i[b,a]+i\sqrt{2}[a,b]\}(a,z)_d$,
7: $\{-\sqrt{2}[b,a]-[a,b]\}(z,a)_s + \{i[b,a]+i\sqrt{2}[a,b]\}(a,z)_d$.

The former represents a wave propagated to the left and the latter to the right, with respect to the phase reference level $x = a$. Hence

$$r = -\frac{\sqrt{2}[b,a]+[a,b]}{i\{[b,a]+\sqrt{2}[a,b]\}} = i\frac{\sqrt{2}+[a,b]^2}{1+\sqrt{2}[a,b]^2}.$$

Evidently $|r| = 1$, since the modulus of an expression of the form $(\sqrt{2}+e^{i\theta})/(1+e^{i\theta}\sqrt{2})$ is unity when θ is real.

This result is correct even when $a = b$, namely when $[a,b] = 1$, in which case a third-order transition point occurs at $z = a$. The use of the Stokes constant $T = 2i \cos\frac{1}{5}\pi$ (see section 4.10) will yield the result

$$r = T + \frac{1}{T} = i\frac{4\cos^2\frac{1}{5}\pi - 1}{2\cos\frac{1}{5}\pi} = i.$$

Example 2. Let $q(x)$ be positive for all x, but possessing double zeros at $z = a$ and at $z = b$. A wave incident from the left is partially transmitted and partially reflected. For $x > b$, let $\arg q = 0$; hence for $a < x < b$, $\arg q = 2\pi$ and for $x < b$, $\arg q = 4\pi$, with the choice of branch cuts shown in Fig. 28. The Stokes constants being $i\sqrt{2}$ for all Stokes lines, it is assumed that the other zeros of q are so placed as not

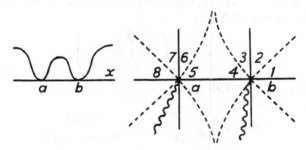

FIG. 27   FIG. 28

to invalidate the uniformity of the approximations throughout the part of the plane depicted.

Then for a wave transmitted to the right for $x > b$, we have

1: $w = (z, b)_d$,
2:       $(z, b)_d + i\sqrt{2}(b, z)_s$,
3:       $(z, b)_s + i\sqrt{2}(b, z)_d$,
4:       $-(z, b)_s + i\sqrt{2}(b, z)_d$,
5:       $-[a, b](z, a)_s + i\sqrt{2}[b, a](a, z)_d$,
6:       $-\{[a, b] + 2[b, a]\}(z, a)_s + i\sqrt{2}[b, a](a, z)_d$,
7:       $-\{[a, b] + 2[b, a]\}(z, a)_d + i\sqrt{2}[b, a](a, z)_s$,
8:       $-\{[a, b] + 2[b, a]\}(z, a)_d - i\sqrt{2}\{[a, b] + [b, a]\}(a, z)_s$,

the former representing the incident wave from the left and the latter the reflected wave.

Dividing by the coefficient of the incident wave, we obtain for the reflection and transmission coefficients respectively

$$r = i\sqrt{2} \frac{[a, b] + [b, a]}{[a, b] + 2[b, a]},$$

$$t = \frac{-1}{[a, b] + 2[b, a]}.$$

A simple calculation shows that $|r|^2 + |t|^2 = 1$, owing to the conservation of energy flow. It has been assumed that the phase-reference level for the waves for $x < a$ is $x = a$, while for the transmitted wave it is $x = b$.

For $a < x < b$, we have $q = e^{2\pi i}|q|$, where $|q|$ is the ordinate exhibited in Fig. 27. Hence

$$[a, b] = \exp\left(-ih \int\limits_a^b |q|^{1/2} dx\right).$$

Writing $\theta = h \int\limits_a^b |q|^{1/2} dx$, we have

$$r = i\sqrt{2} \frac{e^{-i\theta} + e^{i\theta}}{e^{-i\theta} + 2e^{i\theta}} = i\sqrt{2} \frac{2\cos\theta}{3\cos\theta + i\sin\theta},$$

$$t = -1/(3\cos\theta + i\sin\theta);$$

it follows that

$$|r| = \frac{2\sqrt{2}|\cos\theta|}{\sqrt{(8\cos^2\theta+1)}} = \frac{2\sqrt{2}}{\sqrt{(8+\sec^2\theta)}},$$

$$|t| = 1/\sqrt{(8\cos^2\theta+1)}.$$

No reflection takes place when $\cos\theta = 0$, yielding perfect transmission with $|t| = 1$. This occurs when

$$h\int_a^b |q|^{1/2}\,dx = (\tfrac{1}{2}+n)\,\pi,$$

or

$$h = (\tfrac{1}{2}+n)\,\pi \Big/ \int_a^b |q|^{1/2}\,dx \qquad (n\,\text{large}).$$

The medium behaves like a 'window' at these discrete large values of h.

Example 3. Consider the specific profile $q = x^2(a^2-x^2)$; we seek the eigenvalues of h such that evanescent solutions exist both for $x > a$ and for $x < -a$. Transition points of order 1 occur at $x = \pm a$, and a point of order 2 at $x = 0$. An anti-Stokes line exists for $-a < x < a$, and the graph of $q(x)$ and the general disposition of the lines in the complex z-plane are shown in Figs. 29 and 30 respectively.

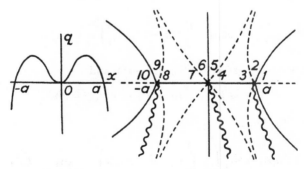

FIG. 29   FIG. 30

In domain 1, we have a subdominant solution $(z, a)_s$, choosing $\arg q = -\pi$ for $x > a$, $\arg q = 0$ for $0 < x < a$, $\arg q = 2\pi$ for $-a < x < a$ and $\arg q = 3\pi$ for $x < -a$. Using i and $i\sqrt{2}$ as the Stokes constants as required, we have

$$
\begin{aligned}
1: \quad & w = (z, a)_s, \\
2: \quad & (z, a)_d, \\
3: \quad & (z, a)_d + i(a, z)_s, \\
4: \quad & [0, a](z, 0)_d + i[a, 0](0, z)_s, \\
5: \quad & [0, a](z, 0)_d + i\{[a, 0] + \sqrt{2}[0, a]\}(0, z)_s, \\
6: \quad & [0, a](z, 0)_s + i\{[a, 0] + \sqrt{2}[0, a]\}(0, z)_d, \\
7: \quad & -\{\sqrt{2}[a, 0] + [0, a]\}(z, 0)_s + i\{[a, 0] + \sqrt{2}[0, a]\}(0, z)_d, \\
8: \quad & -\{\sqrt{2}[a, 0] + [0, a]\}[-a, 0](z, -a)_s + \\
& + i\{[a, 0] + \sqrt{2}[0, a]\}[0, -a](-a, z)_d.
\end{aligned}
$$

In domain 9, there must be no subdominant term, since this would become dominant in 10 and along the negative real axis. Hence

$$
-\{\sqrt{2}[a, 0] + [0, a]\}[-a, 0] - \{[a, 0] + \sqrt{2}[0, a]\}[0, -a] = 0,
$$

being the equation yielding the eigenvalues of h (large).

For $0 < x < a$, $q = x^2(a^2 - x^2)$ and $q^{1/2} = x\sqrt{(a^2 - x^2)}$, so

$$
[0, a] = \exp\left(ih \int_0^a x\sqrt{(a^2 - x^2)}\, dx\right)
$$

$$
= \exp\left(-\frac{ih}{3}\left[(a^2 - x^2)^{3/2}\right]_0^a\right) = \exp\left(\tfrac{1}{3}iha^3\right)
$$

while for $-a < x < 0$, $q = e^{2i\pi}|x^2(a^2 - x^2)|$ and

$$
q^{1/2} = e^{i\pi}|x|\sqrt{(a^2 - x^2)} = x\sqrt{(a^2 - x^2)}
$$

since $x < 0$. Hence

$$
[-a, 0] = \exp\left(ih \int_{-a}^0 x\sqrt{(a^2 - x^2)}\, dx\right) = \exp\left(-\tfrac{1}{3}iha^3\right).
$$

If we now write $e^{i\theta}$ for $\exp(\frac{1}{3}iha^3)$, the equation for h becomes

$$-(\sqrt{2}e^{-i\theta} + e^{i\theta})e^{-i\theta} - (e^{-i\theta} + \sqrt{2}e^{i\theta})e^{i\theta} = 0,$$

reducing to

$$\cos 2\theta = -1/\sqrt{2},$$

or

$$\cos(\tfrac{2}{3}ha^3) = -1/\sqrt{2}.$$

Hence

$$\tfrac{2}{3}ha^3 = \pi \pm \tfrac{1}{4}\pi + 2n\pi,$$

where n is large, positive and integral. The eigenvalues of h are finally given by

$$h = \frac{3\pi}{2a^3}(1 \pm \tfrac{1}{4} + 2n).$$

CHAPTER VI

Applications to Physical Problems

6.1 Schrödinger's equation and the classical limit

We shall consider Schrödinger's equation in its one-dimensional form

$$\frac{d^2\psi}{dx^2} + \frac{8\pi^2 m}{h^2}(H-V)\psi = 0, \tag{6.1}$$

or

$$\psi'' + (8\pi^2 m/h^2)q\psi = 0, \tag{6.2}$$

where $V(x)$ denotes the potential energy and H the total energy of a particle. The use of phase-integral methods enables us to show how the classical limit is attained as $h \to 0$, since $8\pi^2 m/h^2$ may then be treated as a large parameter.

We consider the case of a particle bound in the classical sense between two points $x = a$ and b, such that $H - V(x) > 0$ for $a < x < b$. The boundary conditions imposed are that solutions of (6.2) must be evanescent as $x \to \pm \infty$.

Solutions for this type of equation follow directly from section 5.5. Between the two transition points $x = a$ and b, we have from equation (5.12)

$$\psi = (b,x) - i(x,b),$$

subject to the condition (5.13) that $[a,b]^2 = -1$, yielding the eigenvalues of the energy H. In section 5.13, we proved that

$$\int_{-\infty}^{\infty} \psi\psi^* \, dx = 2\int_a^b q^{-1/2} \, dx = 2\int_a^b (H-V)^{-1/2} \, dx.$$

When ψ is normalized, $\psi\psi^*$ is understood to be the probability density, that is, $\psi\psi^* \, dx$ is the probability that the particle is found in a particular microscopic element dx situated at the position x. Using the unnormalised value of ψ just quoted, we have

$$\psi\psi^* = [(b,x) - i(x,b)][(x,b) + i(b,x)]$$

116

$$\psi\psi^* = 2q^{-1/2} + i(b,x)^2 - i(x,b)^2$$

$$= 2q^{-1/2}\left[1 - \sin\left\{2\left(\frac{8\pi^2 m}{h^2}\right)^{1/2}\int_b^x (H-V)^{1/2}\,dx\right\}\right]$$

$$= 4q^{-1/2}\sin^2\left[\frac{\pi}{4} - \left(\frac{8\pi^2 m}{h^2}\right)^{1/2}\int_b^x (H-V)^{1/2}\,dx\right].$$

Since this fluctuates rapidly when h is small, even in a small macroscopic element δx, we replace the square of the sine by its average value $\frac{1}{2}$, thereby obtaining for the probability that the particle is situated in an element δx:

$$\frac{\text{average } (\psi\psi^*)}{2\int_a^b (H-V)^{-1/2}\,dx}\,\delta x = \frac{q^{-1/2}\,\delta x}{\int_a^b (H-V)^{-1/2}\,dx}.$$

Classically, the energy equation is

$$\tfrac{1}{2}mv^2 + V(x) = H,$$

so $$\delta t = (2/m)(H-V)^{-1/2}\,\delta x.$$

Hence the time that the particle is situated within an element δx divided by the total time for a period is

$$\frac{\delta x}{(H-V)^{1/2}}\bigg/\int_a^b (H-V)^{-1/2}\,dx,$$

identical with the probability given above, thereby supporting the interpretation placed on $\psi\psi^*$. Needless to say, in cases for which the W.K.B.J. solutions are not valid, the predictions of Schrödinger's equation are quite distinct from those of classical dynamics.

The condition that $[a,b]^2 = -1$ implies that

$$\exp\left[2i\left(\frac{8\pi^2 m}{h^2}\right)^{1/2}\int_a^b (H-V)^{1/2}\,dx\right] = -1,$$

or

$$2i\left(\frac{8\pi^2 m}{h^2}\right)^{1/2} \int_a^b (H-V)^{1/2}\,dx \;=\; \pi i + 2\pi n i,$$

where n is an integer. Hence

$$2\sqrt{(2m)} \int_a^b (H-V)^{1/2}\,dx \;=\; (\tfrac{1}{2}+n)\,h,$$

yielding the eigenvalues of the energy H.

But the classical momentum p equals $\pm mv \equiv \pm [2m(H-V)]^{1/2}$, so the equation determining the eigenvalues of H may be written as

$$2\int_a^b p\,dx \;=\; (\tfrac{1}{2}+n)\,h,$$

or

$$\oint p\,dx \;=\; (\tfrac{1}{2}+n)\,h,$$

where the integral is a phase-integral calculated for a complete period.

Historically, the original quantum conditions were expressed in the form

$$\oint p\,dx \;=\; nh, \tag{6.3}$$

such a rule being formulated only on empirical grounds and supported by experiment; no rational interpretation of the rule could be given in terms of classical dynamics. It can therefore be seen that the condition (6.3) represented guesswork founded upon a keen insight into the mathematical representation of the experimental results then available. The fact that differential equations, approximate solutions, evanescent waves and the Stokes phenomenon would be necessary to place the condition on a more satisfactory and logical basis was at that time completely unknown.

6.2 Example using a periodic potential function

The potential function $V(x)$ is given to consist of N equal potential wells $a_1 < x < b_1, \ldots, a_n < x < b_n, \ldots, a_N < x < b_N$, separated by $N-1$

equal potential barriers $b_1 < x < a_2$, ..., $b_n < x < a_{n+1}$, ..., $b_{N-1} < x < a_N$. Find the approximate form of the energy levels.†

Since q is real along the real axis, Jeffreys' connection formulae (4.11) and (4.12) may be used; in particular, formula (4.12) may be used in the reverse sense for reasons discussed in section 4.9. Alternatively, solutions may be traced round the transition points a_n and b_n

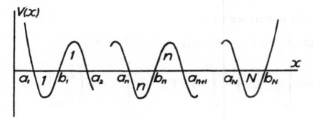

FIG. 31

using the Stokes constant i (or $\frac{1}{2}i$) over the Stokes lines radiating from the transition points of order 1. Branch cuts are inserted as in Fig. 32, in order to preserve the similarity of the wave function to the left of each well.

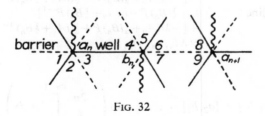

FIG. 32

On the Stokes line $x < a_n$, let the general solution be given by

$$\psi = A_n(a_n, x)_s + B_n(x, a_n)_d$$

† Many other interesting examples of this character are scattered throughout the text, *Problems in Quantum Mechanics*, by I. I. Gol'dman *et al.* [45].

in the sense of section 4.9. Then tracing the solution round, we have

1: $(A_n + \tfrac{1}{2}iB_n)(a_n, z)_s + B_n(z, a_n)_d,$

2: $(A_n + \tfrac{1}{2}iB_n)(a_n, z)_d + B_n(z, a_n)_s,$

3: $(A_n + \tfrac{1}{2}iB_n)(a_n, z)_d + (iA_n + \tfrac{1}{2}B_n)(z, a_n)_s,$

4: $(A_n + \tfrac{1}{2}iB_n)[a_n, b_n](b_n, z)_s + (iA_n + \tfrac{1}{2}B_n)[b_n, a_n](z, b_n)_d$

$\quad \equiv (A_n + \tfrac{1}{2}iB_n)P(b_n, z)_s + (iA_n + \tfrac{1}{2}B_n)P^{-1}(z, b_n)_d,$

where for brevity we write

$$P \equiv [a_1, b_1] = \ldots = [a_n, b_n] = \ldots = [a_N, b_N]$$

$$= \exp\left[i\left(\frac{8\pi^2 m}{h^2}\right)^{1/2} \int\limits_{a_n}^{b_n} (H - V)^{1/2}\, dx\right] \equiv \exp\left(\frac{2\pi i}{h} \int\limits_{a_n}^{b_n} p\, dx\right).$$

Similarly,

5: $[(A_n + \tfrac{1}{2}iB_n)P - i(iA_n + \tfrac{1}{2}B_n)P^{-1}](b_n, z)_s +$

$\quad + (iA_n + \tfrac{1}{2}B_n)P^{-1}(z, b_n)_d,$

6: $[(A_n + \tfrac{1}{2}iB_n)P - i(iA_n + \tfrac{1}{2}B_n)P^{-1}](b_n, z)_d +$

$\quad + (iA_n + \tfrac{1}{2}B_n)P^{-1}(z, b_n)_s,$

Stokes line 6–7: $[(A_n + \tfrac{1}{2}iB_n)P - i(iA_n + \tfrac{1}{2}B_n)P^{-1}](b_n, z)_d +$

$\quad + [-\tfrac{1}{2}i(A_n + \tfrac{1}{2}iB_n)P + \tfrac{1}{2}(iA_n + \tfrac{1}{2}B_n)P^{-1}](z, b_n)_s,$

Stokes line 8–9: $[(A_n + \tfrac{1}{2}iB_n)P - i(iA_n + \tfrac{1}{2}B_n)P^{-1}]S^{-1}(a_{n+1}, z)_s +$

$\quad + [-\tfrac{1}{2}i(A_n + \tfrac{1}{2}iB_n)P + \tfrac{1}{2}(iA_n + \tfrac{1}{2}B_n)P^{-1}] \times$

$\quad \times S(z, a_{n+1})_d,$

where

$$S \equiv [a_{n+1}, b_n] = [a_2, b_1] = \ldots = \exp\left(-\frac{2\pi}{h} \int\limits_{b_n}^{a_{n+1}} |p|\, dx\right) \ll 1.$$

If we write the solution for $x < a_{n+1}$ in the form

$$\psi = A_{n+1}(a_{n+1}, x)_s + B_{n+1}(x, a_{n+1})_d,$$

we see that

$$A_{n+1} = (A_n + \tfrac{1}{2}iB_n)PS^{-1} - i(iA_n + \tfrac{1}{2}B_n)P^{-1}S^{-1},$$

$$B_{n+1} = -\tfrac{1}{2}i(A_n + \tfrac{1}{2}iB_n)PS + \tfrac{1}{2}(iA_n + \tfrac{1}{2}B_n)P^{-1}S,$$

or

$$\begin{pmatrix} A_{n+1} \\ B_{n+1} \end{pmatrix} = \begin{pmatrix} PS^{-1} + P^{-1}S^{-1} & \tfrac{1}{2}iPS^{-1} - \tfrac{1}{2}iP^{-1}S^{-1} \\ -\tfrac{1}{2}iPS + \tfrac{1}{2}iP^{-1}S & \tfrac{1}{4}PS + \tfrac{1}{4}P^{-1}S \end{pmatrix} \begin{pmatrix} A_n \\ B_n \end{pmatrix}$$

$$= \begin{pmatrix} 2e^u \cos v & -e^u \sin v \\ e^{-u} \sin v & \tfrac{1}{2}e^{-u} \cos v \end{pmatrix} \begin{pmatrix} A_n \\ B_n \end{pmatrix}$$

say, where $\quad u = \dfrac{2\pi}{h} \displaystyle\int_{b_n}^{a_{n+1}} |p|\, dx, \qquad v = \dfrac{2\pi}{h} \displaystyle\int_{a_n}^{b_n} p\, dx.$

Hence $\quad \begin{pmatrix} A_{N+1} \\ B_{N+1} \end{pmatrix} = \begin{pmatrix} 2e^u \cos v & -e^u \sin v \\ e^{-u} \sin v & \tfrac{1}{2}e^{-u} \cos v \end{pmatrix}^N \begin{pmatrix} A_1 \\ B_n \end{pmatrix}.$

Evanescent solutions exist for $x < a_1$ if $B_1 = 0$, and for $x > b_N$ if $A_{N+1} = 0$ (that is, by examination of the solution on the Stokes line 6–7 with $n = N$). Hence

$$\begin{pmatrix} 0 \\ B_{N+1} \end{pmatrix} = \begin{pmatrix} 2e^u \cos v & -e^u \sin v \\ e^{-u} \sin v & \tfrac{1}{2}e^{-u} \cos v \end{pmatrix}^N \begin{pmatrix} A_1 \\ 0 \end{pmatrix},$$

implying that the first element of the Nth power of this 2×2 matrix must vanish. This equation provides the condition for the energy levels.

If the 2×2 matrix is abbreviated to

$$\begin{pmatrix} a & b \\ c & d \end{pmatrix},$$

let the two characteristic roots be λ and μ, and the two vectors be

$$\begin{pmatrix} d - \lambda \\ -c \end{pmatrix} \quad \text{and} \quad \begin{pmatrix} d - \mu \\ -c \end{pmatrix}$$

respectively. Then elementary matrix theory demonstrates that

$$\begin{pmatrix} a & b \\ c & d \end{pmatrix} = \begin{pmatrix} d - \lambda & d - \mu \\ -c & -c \end{pmatrix}^{-1} \begin{pmatrix} \lambda & 0 \\ 0 & \mu \end{pmatrix} \begin{pmatrix} d - \lambda & d - \mu \\ -c & -c \end{pmatrix},$$

so
$$\begin{pmatrix} a & b \\ c & d \end{pmatrix}^N = \begin{pmatrix} d-\lambda & d-\mu \\ -c & -c \end{pmatrix}^{-1} \begin{pmatrix} \lambda^N & 0 \\ 0 & \mu^N \end{pmatrix} \begin{pmatrix} d-\lambda & d-\mu \\ -c & -c \end{pmatrix},$$

the first element being proportional to

$$(-c \quad -d+\mu)\begin{pmatrix} \lambda^N & 0 \\ 0 & \mu^N \end{pmatrix}\begin{pmatrix} d-\lambda \\ -c \end{pmatrix},$$

so
$$-c\lambda^N(d-\lambda) - c\mu^N(-d+\mu) = 0$$

or
$$(\lambda^N - \mu^N)d = \lambda^{N+1} - \mu^{N+1}. \tag{6.4}$$

The characteristic equation for the 2×2 matrix is

$$\begin{vmatrix} a-\lambda & b \\ c & d-\lambda \end{vmatrix} = 0,$$

or
$$\lambda^2 - (2e^u\cos v + \tfrac{1}{2}e^{-u}\cos v)\lambda + 1 = 0,$$

which, if we write

$$2e^u\cos v + \tfrac{1}{2}e^{-u}\cos v = 2\cos\phi, \tag{6.5}$$

has the roots $\lambda, \mu = e^{\pm i\phi}$.

Equation (6.4) now reduces to

$$\tfrac{1}{2}e^{-u}\cos v\sin N\phi = \sin(N+1)\phi. \tag{6.6}$$

Eliminating $\cos v$ from equations (6.5) and (6.6), we obtain after elementary simplification

$$4e^{2u} = \sin(N-1)\phi/\sin(N+1)\phi.$$

Since e^{2u} is large, this equation is solved approximately by placing

$$\sin(N+1)\phi = 0,$$

provided $\sin(N-1)\phi \neq 0$. Hence

$$\phi = n\pi/(N+1),$$

where n is an integer, provided $\phi = 0, \pi, 2\pi, \ldots$ are excluded. Hence there are N distinct values of $\cos\phi$ given by $\cos[n\pi/(N+1)]$, where $n = 1, 2, \ldots, N$.

Equation (6.5) now becomes

$$(2e^u + \tfrac{1}{2}e^{-u})\cos v = 2\cos[n\pi/(N+1)],$$

or

$$\cos v \doteq e^{-u}\cos[n\pi(N+1)],$$

so

$$v = \pi(m+\tfrac{1}{2}) - (-1)^m \sin^{-1}\left[e^{-u}\cos\frac{n\pi}{N+1}\right]$$

$$= \pi(m+\tfrac{1}{2}) - (-1)^m e^{-u}\cos\frac{n\pi}{N+1}$$

where $m = 0, 1, 2, 3, \ldots$ and $n = 1, 2, \ldots, N$. In terms of the phase-integrals, the equation giving the energy levels is

$$\frac{2\pi}{h}\int_{a_n}^{b_n} p\,dx = \pi(m+\tfrac{1}{2}) - (-1)^m \exp\left(-\frac{2\pi}{h}\int_{b_n}^{a_{n+1}} |p|\,dx\right)\cos\frac{n\pi}{N+1}.$$

Apart from the last factor, this result is similar to the condition for a separate well. The effect of the last term is to split each level into N sublevels. The sign $-(-1)^m$ may be replaced by $+$, since the term $\cos[n\pi/(N+1)]$ produces a $\pm$ for each sublevel for each m.

6.3 The bounded harmonic oscillator

If we write $V(x) = \tfrac{1}{2}kx^2$, and use the boundary condition that $\psi \to 0$ as $x \to \pm\infty$, we have the case of the harmonic oscillator. From section 5.5, we see that the exact equation giving the energy levels is

$$[-a, +a]^2 = -1,$$

where $\pm a$ denote the zeros of $H - \tfrac{1}{2}kx^2$. It follows that

$$\frac{2\pi\sqrt{(2m)}}{h}\int_{-a}^{+a} \sqrt{(H - \tfrac{1}{2}kx^2)}\,dx = (n+\tfrac{1}{2})\pi,$$

or

$$\frac{2\pi}{h}\left(\frac{m}{k}\right)^{1/2} H = n+\tfrac{1}{2},$$

the integral itself being merely proportional to the area of a semi-circle, namely $\sqrt{(\frac{1}{2}k)}\frac{1}{2}\pi a^2$, where $H = \frac{1}{2}ka^2$.

The classical equation of motion for the particle is

$$m\ddot{x} = -d(\tfrac{1}{2}kx^2)/dx = -kx,$$

so the frequency of the vibrations is given by $\nu = (k/m)^{1/2}/2\pi$. Eliminating $(m/k)^{1/2}$, we find that

$$H = (n + \tfrac{1}{2})\,h\nu,$$

giving the discrete set of allowed energy levels for the harmonic oscillator.

If now, the condition at infinity is replaced by the condition that ψ vanishes at the walls of an enclosure $x = \pm b$, we may obtain an equation for the energy levels of a bounded harmonic oscillator. In Fig. 33, the walls $\pm b$ are fixed, but the transition points $\pm a$ are

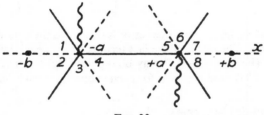

FIG. 33

discretely variable depending upon the particular energy level taken. We restrict ourselves to the case when the W.K.B.J. solutions are valid in the ranges $-b < x < -a$, $-a < x < +a$, $+a < x < +b$.

The general solution on the anti-Stokes line joining the two transition points may be written in the form

$$\psi = A(x, a) + B(a, x),$$

the branch of the square root being chosen so that (x, a) represents a wave to the right. Tracing the solution round, we obtain

5: $A(z, a)_d + B(a, z)_s,$
6: $A(z, a)_d + (B - iA)(a, z)_s,$
7: $A(z, a)_s + (B - iA)(a, z)_d,$

Stokes line 7–8: $\frac{1}{2}(A-iB)(x,a)_s + (B-iA)(a,x)_d$,

using the principles outlined in section 4.9. Similarly,

4: $A[-a,a](z,-a)_s + B[a,-a](-a,z)_d$,

3: $\{A[-a,a]-iB[a,-a]\}(z,-a)_s + B[a,-a](-a,z)_d$,

2: $\{A[-a,a]-iB[a,-a]\}(z,-a)_d + B[a,-a](-a,z)_s$,

Stokes line 1–2: $\{A[-a,a]-iB[a,-a]\}(x,-a)_d +$

$$+\{-\tfrac{1}{2}iA[-a,a]+\tfrac{1}{2}B[a,-a]\}(-a,x)_s.$$

Now $\psi = 0$ when $x = \pm b$; hence

$$\tfrac{1}{2}(A-iB)(b,a)_s + (B-iA)(a,b)_d = 0,$$

$$\{A[-a,a]-iB[a,-a]\}(-b,-a)_d +$$
$$+\{-\tfrac{1}{2}iA[-a,a]+\tfrac{1}{2}B[a,-a]\}(-a,b)_s = 0.$$

Let $\arg q = -\pi$, 0, $-\pi$ respectively for the three ranges $x > a$, $-a < x < a$, $x < -a$, and we write

$$[a,b]_d = [-b,-a]_d = L, \qquad [-a,a] = P;$$

then

$$\tfrac{1}{2}(A-iB)L^{-1} + (B-iA)L = 0,$$

$$(AP - iBP^{-1})L + (-\tfrac{1}{2}iAP + \tfrac{1}{2}BP^{-1})L^{-1} = 0.$$

The determinant that eliminates A and B is

$$\begin{vmatrix} \tfrac{1}{2}L^{-1}-iL & -\tfrac{1}{2}iL^{-1}+L \\ PL-\tfrac{1}{2}iPL^{-1} & -iP^{-1}L+\tfrac{1}{2}P^{-1}L^{-1} \end{vmatrix} = 0;$$

upon simplification, this reduces to

$$P^2 = -\left(\frac{1+\tfrac{1}{2}iL^{-2}}{1-\tfrac{1}{2}iL^{-2}}\right)^2 = -1-2iL^{-2}$$

to the second order in L^{-2}.

Now

$$P^2 = \exp\left[2.\frac{2\pi i\sqrt{(2m)}}{h}\int_{-a}^{+a}\sqrt{(H-\tfrac{1}{2}kx^2)}\,dx\right] = \exp(2\pi iH/h\nu)$$

as before. Moreover,

$$L^{-2} = \exp\left[-2.\frac{2\pi\sqrt{(2m)}}{h} \int_a^b \sqrt{(\tfrac{1}{2}kx^2 - H)}\,dx \right]$$

$$= \exp\left[-\frac{4\pi\sqrt{(mk)}}{h}\left\{ \tfrac{1}{2}b\sqrt{\left(b^2 - \frac{2H}{k} \right)} - \right.\right.$$
$$\left.\left. -\frac{H}{k}\log\frac{b + \sqrt{(b^2 - 2H/k)}}{\sqrt{(2H/k)}} \right\} \right]$$

upon integration by parts. If $b^2 \gg 2H/k$, we obtain

$$L^{-2} = \exp\left[-\frac{2\pi\sqrt{(mk)}}{h}\left(b^2 - \frac{H}{k} - \frac{H}{k}\log\frac{2b^2 k}{H} \right) \right]$$

$$= \exp\left[-2\pi\sqrt{(mk)}\,b^2/h \right]\exp\left(H/h\nu \right)(2b^2 k/H)^{H/h\nu}$$

upon simplification.

The equation determining H now becomes

$$\exp\left(2\pi i H/h\nu \right) = -1 - 2i\exp\left[-2\pi\sqrt{(mk)}\,b^2/h \right] \times$$
$$\times \exp\left(H/h\nu \right)(2b^2 k/H)^{H/h\nu}.$$

A first approximation is obviously

$$2\pi H/h\nu = (2n+1)\,\pi,$$

or
$$H = (n + \tfrac{1}{2})\,h\nu.$$

If the second approximation is given by $H = (n + \tfrac{1}{2})\,h\nu + \epsilon$, we have

$$\exp\left\{ 2\pi i[(n + \tfrac{1}{2}) + \epsilon/h\nu] \right\} = -\exp\left(2\pi i\epsilon/h\nu \right) \doteq -(1 + 2\pi i\epsilon/h\nu),$$

so $\quad \epsilon = (h\nu/\pi)\exp\left[-2\pi\sqrt{(mk)}\,b^2/h \right]\exp\left(H/h\nu \right)(2b^2 k/H)^{H/h\nu},$

in which result H is placed equal to its first approximation. Hence

$$H = (n + \tfrac{1}{2})\,h\nu + (h\nu/\pi)\exp\left[-2\pi\sqrt{(mk)}\,b^2/h \right] \times$$
$$\times \exp\left(n + \tfrac{1}{2} \right)\,[2b^2 k/(n + \tfrac{1}{2})h\nu]^{n + 1/2}.$$

A necessary condition for the validity of the W.K.B.J. solutions between the transition points is given by inequality (2.8). For the present problem, h^2 is replaced by $8\pi^2 m/h^2$ and q by $H - \frac{1}{2}kx^2$, yielding

$$\frac{8\pi^2 m}{h^2} \gg \left| \frac{5}{16} \cdot \frac{k^2 x^2}{(H - \frac{1}{2}kx^2)^3} + \frac{k}{4(H - \frac{1}{2}kx^2)^2} \right|.$$

When $x = 0$, this reduces to $2\sqrt{2}\,(n + \frac{1}{2}) \gg 1$, showing that high energy levels alone are allowed for the validity of the calculation.

This result is in agreement with that derived by Auluck and Kothari [4] by another method, subject to their formula (24) being evaluated when n (their q) is large. However, their correction term is double that which we have produced. The reason is that a factor 2 was accidently dropped in their paper † when quoting the asymptotic expressions for the parabolic cylinder functions, for which the coefficient of the subdominant term was required in the presence of a dominant term. The averaging technique proved in section 4.9 was not appreciated when their paper was written.

6.4 The hydrogen atom
Schrödinger's equation in three dimensions for the hydrogen atom is given by

$$\nabla^2 \psi + (8\pi^2 m/h^2)(H + e^2/r)\psi = 0,$$

where $\nabla^2 \psi$ denotes the Laplacian in spherical polar coordinates. The potential energy for the system consisting of the central nucleus and the one orbital electron is $-e^2/r$, where $-e$ is the electronic charge.

Any text-book on wave mechanics will show that separable solutions may be found in terms of the three spherical polar coordinates r, θ, ϕ, the ordinary differential equation for the radial wave-function $R(r)$ being

$$R'' + \frac{2}{r} R' + \left[\frac{8\pi^2 m}{h^2}\left(H + \frac{e^2}{r} \right) - \frac{l(l+1)}{r^2} \right] R = 0.$$

† The author thanks Professor F. C. Auluck of Delhi for informing him that this error has been pointed out by Singh and Baijal [119], Hull and Julius [120] and Singh [121].

Here, $l = 0, 1, 2, 3, \ldots$, these integers arising from the solution of the equation involving θ only. The eigenvalues of the energy H are to be found from this equation for the radial wave-function, the boundary conditions being that $R \to 0$ as $r \to +\infty$ and that R remains finite as $r \to 0$.

Writing $R = r^{-1} S$, we find

$$R' = r^{-1} S' - r^{-2} S,$$
$$R'' = r^{-1} S'' - 2r^{-2} S' + 2r^{-3} S;$$

this transformation eliminates the term containing the first derivative, yielding the equation

$$S'' + [(8\pi^2 m/h^2)(H + e^2/r) - l(l+1)/r^2] S = 0.$$

We require $S \to 0$ as $r \to +\infty$ and as $r \to 0$.

This equation resembles the form

$$S'' + (8\pi^2 m/h^2) q S = 0,$$

with
$$q = H + e^2/r - h^2 l(l+1)/(8\pi^2 m r^2)$$
$$= [Hr^2 + e^2 r - h^2 l(l+1)/8\pi^2 m] r^{-2}.$$

For an evanescent solution along the Stokes line $r \to +\infty$, we obviously require $H < 0$ so $q < 0$; let us write $H = -W$.

This form of q has two zeros r_1 and r_2 $(r_1 < r_2)$ given by

$$r_1, r_2 = (2W)^{-1}\{e^2 \pm \sqrt{[e^4 - Wh^2 l(l+1)/2\pi^2 m]}\} > 0.$$

It appears then that we have an underdense region between the two zeros, and that solutions that tend to zero are required on each side of the underdense region. The eigenvalues of W would then be given by the equation $[r_1, r_2]^2 = -1$.

Unfortunately, the eigensolutions for ψ calculated under these assumptions are not identical with the asymptotic forms of the exact solutions found by other mathematical techniques. To overcome this difficulty, many writers have merely replaced $l(l+1)$ by $(l+\frac{1}{2})^2$ without discussion; Langer [75] (at the bottom of page 674) has quoted six references to such authors. Even in more recent times, Landau and

Lifshitz [73] (page 169) make this replacement merely by comparing the W.K.B.J. solutions with the exact ones. This is a particularly weak method of approach, suggesting that the W.K.B.J. method is incapable of producing results that should rightly be expected of it without comparing it with results already known more accurately by other considerations. However, Langer [75] has shown that the fault lies not with the W.K.B.J. method but with the application of it. The reason is that the W.K.B.J. solution for $r < r_1$ cannot be valid *up to* $r = 0$, at which point a condition must be applied to S. In fact, a sub-dominant solution to the left of r_1 would become infinite at $r = 0$. Langer has shown how to avoid this difficulty.

Let $r = e^x$ and $R = e^{-x/2} U$ in the equation for R, thereby extending the range of the independent variable from $-\infty$ to $+\infty$. We have

$$\frac{dR}{dr} = e^{-x}\frac{dR}{dx},$$

$$\frac{d^2R}{dr^2} = e^{-2x}\frac{d^2R}{dx^2} - e^{-2x}\frac{dR}{dx},$$

$$\frac{dR}{dx} = e^{-x/2}\frac{dU}{dx} - \tfrac{1}{2}e^{-x/2}U,$$

$$\frac{d^2R}{dx^2} = e^{-x/2}\frac{d^2U}{dx^2} - e^{-x/2}\frac{dU}{dx} + \tfrac{1}{4}e^{-x/2}U.$$

Substitution and simplification yield

$$U'' + [(8\pi^2 m/h^2)(He^{2x} + e^2 e^x) - (l + \tfrac{1}{2})^2]\,U = 0,$$

an equation for which the W.K.B.J. solutions are valid up to $x = -\infty$ (that is, $r = 0$). An extra $\tfrac{1}{4}$ has been added to $l(l+1)$, this extra term arising from the transformations of the derivatives involved.

The zeros of

$$q = -We^{2x} + e^2 e^x - h^2(l + \tfrac{1}{2})^2/8\pi^2 m$$

are given by the values x_1, x_2 specified by

$$e^x = (2W)^{-1}\{-e^2 \pm \sqrt{[e^4 - Wh^2(l + \tfrac{1}{2})^2/2\pi^2 m]}\}.$$

Hence the eigenvalues of W necessitated by subdominant solutions at $\pm \infty$ are given by the equation $[x_1, x_2]^2 = -1$, namely by

$$2\left(\frac{8\pi^2 mW}{h^2}\right)^{1/2} \int_{x_1}^{x_2} \left(-e^{2x} + \frac{e^2}{W}e^x - \frac{(l+\frac{1}{2})^2 h^2}{8\pi^2 mW}\right)^{1/2} dx = (2n+1)\pi.$$

If $y = e^x$, we obtain

$$\left(\frac{8\pi^2 mW}{h^2}\right)^{1/2} \int_{y_1}^{y_2} \left(-y^2 + \frac{e^2}{W}y - \frac{(l+\frac{1}{2})^2 h^2}{8\pi^2 mW}\right)^{1/2} \frac{dy}{y} = (n+\frac{1}{2})\pi.$$

An eigenvalue problem concerning the thermosphere (an atmosphere whose temperature increases without limit as the altitude increases) that necessitates the evaluation of a similar phase-integral has been considered by Eckart [29] (page 214).

Now any comprehensive set of tables of indefinite integrals provides the value

$$\int \frac{\sqrt{(-ay^2 + by - c)}\, dy}{y} = \quad \sqrt{(-ay^2 + by - c)} -$$

$$- \frac{b}{2\sqrt{a}}\sin^{-1}\frac{b - 2ay}{\sqrt{(b^2 - 4ac)}} -$$

$$- \sqrt{c}\sin^{-1}\frac{by - 2c}{y\sqrt{(b^2 - 4ac)}},$$

where a, b, $c > 0$, and where y lies between the two real zeros of the expression $-ay^2 + by - c$. The inverse sines are restricted to lie in the range $-\frac{1}{2}\pi$ to $\frac{1}{2}\pi$. If the limits are specified by

$$y_1, y_2 = [b \pm \sqrt{(b^2 - 4ac)}]/2a,$$

the values of the integral at the upper and lower limits reduce simply to

$$-(b/2\sqrt{a})\sin^{-1}(\mp 1) - \sqrt{c}\sin^{-1}(\pm 1),$$

so the definite integral has the value

$$(b/2\sqrt{a} - \sqrt{c})\pi.$$

Substituting $a = 1$, $b = e^2/W$, $c = (l+\tfrac{1}{2})^2 h^2/8\pi^2\, mW$, we obtain

$$\left(\frac{8\pi^2\, mW}{h^2}\right)^{1/2}\left(\frac{e^2}{2W} - \frac{(l+\tfrac{1}{2})h}{(8\pi^2\, mW)^{1/2}}\right) = n+\tfrac{1}{2},$$

or

$$\left(\frac{8\pi^2\, mW}{h^2}\right)^{1/2}\frac{e^2}{2W} = (n+\tfrac{1}{2})+(l+\tfrac{1}{2})$$

$$= \text{an integer } p.$$

Hence

$$W = -H = 2\pi^2\, me^4/h^2 p^2.$$

This result has been proved only when the W.K.B.J. solutions are valid between the two transition points, but the exact theory for the hydrogen atom shows that the energy levels are always of this form. The form of the radial wave-function $R(r)$ under these assumptions is given in Landau and Lifshitz [73] (page 170) and may be consulted if required.

6.5 Bremmer's interpretation of the W.K.B.J. solutions
To the left of $x = 0$ let $q \equiv 1$, while to the right of $x = 0$ let the medium be subdivided into small sections or layers by the points $x_1, x_2, \ldots,$ $x_r, \ldots, x_n$. Let $q = q_r$, a constant, throughout the range $x_{r-1} < x < x_r$, where the values of q vary but slowly from section to section.

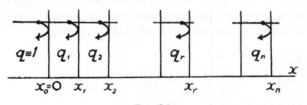

FIG. 34

Bremmer [20] considered a primary wave incident from the left on to $x = 0$. At the interface $x = 0$, at which w and w' are continuous, a small fraction of the incident wave is reflected, but the larger portion of the wave is transmitted. Similarly at $x = x_1$, a second portion is reflected while the larger portion is transmitted. The transmitted wave, disregarding all reflected and multiply-reflected waves, is traced through the medium, its value being found at $x = x_n$; the limit is then considered as the width of the sections tends to zero thereby producing a continuously variable medium.

For the section just to the left of x_r, let

$$w = A_r e^{-ihQ_r x} + B_r e^{ihQ_r x},$$

and just to the right let

$$w = C_r e^{-ihQ_{r+1} x},$$

where $Q_r = \sqrt{q_r}$, the branch having been chosen so that negative signs refer to waves propagated to the right. The continuity of w and w' over the boundary $x = x_r$ yields the two equations

$$A_r e^{ihQ_r x_r} + B_r e^{ihQ_r x_r} = C_r e^{-ihQ_{r+1} x_r},$$

$$-Q_r A_r e^{-ihQ_r x_r} + Q_r B_r e^{ihQ_r x_r} = -Q_{r+1} C_r e^{-ihQ_{r+1} x_r}.$$

Eliminating B_r, we obtain

$$C_r = 2 A_r Q_r (Q_r + Q_{r+1})^{-1} e^{-ih(Q_r - Q_{r+1}) x_r},$$

where $Q_0 = 1$, $x_0 = 0$ and $A_0 = 1$ if the incident wave equals e^{-ihx}.

Hence when $x = x_n$, the transmitted wave under these conditions is given by using this formula for $r = 0, 1, 2, \ldots, n$, yielding

$$w_0 = C_{n+1} e^{-ihQ_{n+1} x_n}$$

$$= \frac{2Q_0}{Q_0 + Q_1} \cdot \frac{2Q_1}{Q_1 + Q_2} \cdot \ldots \cdot \frac{2Q_n}{Q_n + Q_{n+1}} \times$$

$$\times e^{-ih(Q_0 - Q_1) x_0} e^{-ih(Q_1 - Q_2) x_1} \ldots e^{-ih(Q_n - Q_{n+1}) x_n} e^{-ihQ_{n+1} x_n}.$$

Now

$$\frac{2Q_0}{Q_0 + Q_1} = \frac{2Q_0}{2Q_0 + (Q_1 - Q_0)} \doteq 1 - \frac{Q_1 - Q_0}{2Q_0} \doteq \exp\left(-\frac{Q_1 - Q_0}{2Q_0}\right),$$

so

$$w_0 = \exp\left(-\frac{Q_1-Q_0}{2Q_0} - \frac{Q_2-Q_1}{2Q_1} - \ldots - \frac{Q_{n+1}-Q_n}{2Q_n}\right) \times$$

$$\times \exp\left[-ihQ_0x_0 - ihQ_1(x_1-x_0) - \ldots - ihQ_n(x_n-x_{n-1})\right]$$

$$= \exp\left(-\sum\frac{\delta Q}{2Q} - ih\sum Q\delta x\right).$$

In the limit, we may formally replace the summation by integration, giving

$$w_0 = \exp\left(-\int_0^x \frac{dQ}{2Q} - ih\int_0^x Q\,dx\right)$$

$$= Q^{-1/2}\exp\left(-ih\int_0^x Q\,dx\right)$$

$$= q^{-1/4}\exp\left(-ih\int_0^x \sqrt{q}\,dx\right),$$

representing the W.K.B.J. solution corresponding to a wave propagated to the right.

Hence this particular W.K.B.J. solution originates by discarding all reflections from the incident wave; all multiple reflections that would yield a further wave propagated to the right are also neglected. Evidently such considerations must of necessity exclude a range of x in which $q = 0$.

Bremmer has further shown that the wave formed by single reflection processes at all the boundaries may be taken as a first correction term; his analysis shows that this additional wave propagated to the left is given by

$$w_1(x) = -\tfrac{1}{4}q^{-1/4}(x)\int_x^\infty ds\, q'(s)q^{-3/4}(s)\,w_0(s)\exp\left(ih\int_x^s \sqrt{q}(\sigma)\,d\sigma\right).$$

Successive approximations to this series have been considered by

Bellman and Kalaba [8, 9], who examined the convergence of the series and showed that the sum actually satisfies the original wave equation.

6.6 Ray theory in an isotropic ionosphere

The propagation of electromagnetic waves in an ionized medium is influenced by the presence of free electrons, by the collisions of these electrons with heavier particles causing energy losses and by the existence of the earth's magnetic field which renders the ionosphere *anisotropic* or *doubly-refracting* with two refractive indices instead of one. Here, we consider only the effect of the ionisation density. The complete theory for the anisotropic ionosphere may be found in Budden [24] (Chapter XIII) and a similar though less comprehensive treatment for the hydrodynamics of stratified atmospheres and oceans in Eckart [29] (Chapter XI). The work of Booker [14] should also be mentioned in this connection.

Maxwell's equations modified by the existence of a current density **j** caused by the motion of the free electrons are given by

$$\text{curl}\,\mathbf{E} = -\partial\mathbf{H}/c\,\partial t, \qquad \text{curl}\,\mathbf{H} = \partial\mathbf{E}/c\,\partial t + 4\pi\mathbf{j}/c.$$

If all field quantities contain the time factor e^{ipt}, then

$$\text{curl}\,\mathbf{E} = -ik\mathbf{H}, \qquad \text{curl}\,\mathbf{H} = ik\mathbf{E} + 4\pi\mathbf{j}/c,$$

where $k = p/c$. Eliminating **H**, we obtain

$$\text{curl}\,\text{curl}\,\mathbf{E} = k^2\mathbf{E} - 4\pi ik\mathbf{j}/c.$$

If **r** is the vector displacement of a volume element δV from its equilibrium position, and if N and m denote the electron density and the electronic mass respectively, the equation of motion of the element under the action of the electric field alone is

$$Nm\delta V\,d^2\mathbf{r}/dt^2 = -Ne\delta V\mathbf{E},$$

or

$$mp^2\mathbf{r} = e\mathbf{E}.$$

The current density **j** is now given by

$$\mathbf{j} = \rho\,d\mathbf{r}/dt = -Neip\mathbf{r} = Neip(e\mathbf{E}/mp^2).$$

Hence $\qquad$ $\operatorname{curl} \operatorname{curl} \mathbf{E} = \left(k^2 - \frac{4\pi k}{c} \cdot \frac{Ne^2 p}{mp^2} \right) \mathbf{E}$

$$= k^2 (1 - X) \mathbf{E}, \tag{6.7}$$

where $X = 4\pi Ne^2/mp^2 \equiv k_0^2/k^2$, say. This equation is the differential equation for the electric field in the ionized medium.

Consider the simple case of horizontal stratification in a coordinate system in which Oz is vertical, Ox and Oy forming a horizontal plane. N is a given function of the height z only, such that $N = 0$ for $z < 0$, the Ox, y-plane forming the base of the ionization. We prescribe an x variation for all field quantities of the form $e^{-ik \sin \theta x}$, where $\sin \theta$ remains to be interpreted; we assume for simplicity that no variation with respect to y occurs in all field variables.

If a prime denotes differentiation with respect to z, we have

$$\operatorname{curl} \operatorname{curl} \mathbf{E} = \operatorname{curl} \begin{pmatrix} -ik \sin \theta & 0 & ' \\ E_x & E_y & E_z \end{pmatrix}$$

$$= \begin{pmatrix} -ik \sin \theta & 0 & ' \\ -E_y' & ik \sin \theta E_z + E_x' & -ik \sin \theta E_y \end{pmatrix}.$$

The y-component is $k^2 \sin^2 \theta E_y - E_y''$, so the y-component of equation (6.7) is

$$k^2 \sin^2 \theta E_y - E_y'' = k^2 (1 - X) E_y,$$

or $\qquad$ $E_y'' + k^2 (\cos^2 \theta - X) E_y = 0,$

an equation of the type for which the W.K.B.J. method is applicable for high enough frequencies.

In free space below $z = 0$, we have $X = 0$, so $E_y = e^{\pm ik \cos \theta z}$, apart from the x and t variations. Totally,

$$E_y = \exp [-ik(x \sin \theta \pm z \cos \theta) + ipt],$$

representing a plane wave incident at an angle θ to the vertical or reflected at an angle θ to the vertical. This provides the interpretation of the angle θ.

Within the medium we may write

$$E'' + k^2 qE = 0,$$

where the suffix y is dropped and where $\cos^2\theta - X$ is replaced by q. A less exact form of the W.K.B.J. solutions is generally used in the context of ray theory. Assume that

$$E = \exp\{ik[ct - x\sin\theta - \phi(z)]\},$$

where $\phi(z) = \cos\phi z$ in free space. The less accurate form of the W.K.B.J. solution is obtained by differentiating only the exponential term and not any other factor that may multiply it. Then

$$E' = -ik\phi'(z)\exp\{ik[ct - x\sin\theta - \phi(z)]\},$$

$$E'' \rightleftharpoons -k^2(\phi')^2\exp\{ik[ct - x\sin\theta - \theta(z)]\}.$$

The equation for E becomes

$$-k^2(\phi')^2 + k^2 q = 0.$$

This is the simplified version of the equation that may be termed the *Hamilton-Jacobi* equation associated with the field equation. Evidently

$$\phi = \int_0^z \sqrt{q}\,dz,$$

typical of the first approximation to the Ricatti equation considered in section 2.1. We use then the approximate solution

$$E = \exp\left[ik\left(ct - Sx - \int_0^z \sqrt{q}\,dz\right)\right] \equiv \exp(iP),$$

say, where $S \equiv \sin\theta$.

Consider the integrated superposition E_s of such waves for varying S and k, with an amplitude factor $A(S,k)$ confined to a narrow maximum near a particular value of k and S; such will represent a *pulse* or *wave-packet*:

$$E_s = \iint A\exp(iP)\,dS\,dk.$$

For any given value of x, z, t, the phase P as S and k vary will be such that destructive interference takes place yielding but little contribution to the integral; in fact, the contribution to the integral will be appreciable only for those values of S and k for which the phase P is stationary. Then

$$\partial P/\partial S \equiv k(-x - \partial\phi/\partial S) = 0,$$
$$\partial P/\partial k \equiv ct - Sx - \phi - k\,\partial\phi/\partial k = 0.$$

The path of the pulse or the ray is given by the first equation:

$$x = -\frac{\partial\phi}{\partial S} = -\frac{\partial}{\partial S}\int_0^z \sqrt{q}\,dz$$

$$= -\frac{\partial}{\partial S}\int_0^z \sqrt{(1 - S^2 - X)}\,dz$$

$$= S\int_0^z \frac{dz}{\sqrt{q}}. \tag{6.8}$$

Two points (x, z), $(x + \delta x, z + \delta z)$ on the path are connected to the time differential δt by

$$\delta x + \delta(\partial\phi/\partial S) = 0,$$
$$c\,\delta t = S\,\delta x + \delta\phi + k\,\delta(\partial\phi/\partial k),$$

where

$$\delta\frac{\partial\phi}{\partial S} = \frac{\partial}{\partial S}(\sqrt{q})\,\delta z = \frac{\partial}{\partial S}\sqrt{(1 - S^2 - X)}\,\delta z = -\frac{S}{\sqrt{q}}\,\delta z,$$

and

$$\delta\frac{\partial\phi}{\partial k} = \frac{\partial}{\partial k}(\sqrt{q})\,\delta z = \frac{\partial}{\partial k}\sqrt{(\cos^2\theta - k_0^2/k^2)}\,\delta z = \frac{k_0^2/k^3}{\sqrt{q}}\,\delta z,$$

yielding

$$\delta x = (S/\sqrt{q})\,\delta z, \tag{6.9}$$

$$c\,\delta t = S\,\delta x + \sqrt{(\cos^2\theta - X)}\,\delta z + \frac{(k_0^2/k^2)\,\delta z}{\sqrt{(\cos^2\theta - X)}}$$

$$= S\,\delta x + \frac{\cos^2\theta\,\delta z}{\sqrt{(\cos^2\theta - X)}}. \tag{6.10}$$

The velocity of the wave-packet is known as the *group velocity* v. Its value is given by

$$v^2 = \left(\frac{dx}{dt}\right)^2 + \left(\frac{dz}{dt}\right)^2 = \left(\frac{S^2}{q}+1\right)\left(\frac{dz}{dt}\right)^2$$

from equation (6.9). Further, when δx is eliminated between equations (6.9) and (6.10), we obtain

$$c\,\delta t = (S^2\,\delta z + \cos^2\theta\,\delta z)/\sqrt{q} = \delta z/\sqrt{q},$$

so
$$v^2 = (S^2/q+1)\,c^2 q = c^2(S^2+q) = c^2(1-X),$$

and
$$v = c\sqrt{(1-X)},$$

which is always real since $\cos^2\theta - X \geqslant 0$.

Equation (6.9) provides the gradient of the ray, namely

$$dz/dx = q^{1/2}/\sin\theta. \tag{6.11}$$

This vanishes when $q = 0$, namely when $\cos^2\theta = X$, yielding the maximum value of z attained by the ray path. In this interpretation, it should be recalled that the W.K.B.J. approximations break down when $q = 0$; in fact the W.K.B.J. solutions do not define the concept of a ray in the immediate neighbourhood of the height where $q = 0$.

As an example, consider a linear increase in electron density given by $X = az/k^2$ say. The ray path (6.8) is given by

$$x = \sin\theta \int_0^z \frac{dz}{\sqrt{(\cos^2\theta - az/k^2)}}$$

$$= (2k^2\sin\theta/a)[\cos\theta - \sqrt{(\cos^2\theta - az/k^2)}],$$

a parabola, whose vertex is at the height $k^2\cos^2\theta/a$.

The *complex Poynting vector* governing the flow of energy is proportional to $\mathbf{E}^* \wedge \mathbf{H} \propto \mathbf{E}^* \wedge \operatorname{curl}\mathbf{E}$; this equals

$$(0, E_y^*, 0) \wedge \operatorname{curl}(0, E_y, 0) = (-ik\sin\theta E_{y}^{\,!}E_y^*, 0, E_y^* E_y').$$

Its direction is specified by the gradient

$$\frac{E^* E_y'}{-ik\sin\theta\, E_y E_y^*} = \frac{-ik\,\phi'}{-ik\sin\theta} = \frac{\sqrt{q}}{\sin\theta}$$

identical with (6.11), showing that the flow of energy takes place along the ray path.

6.7 Generalizations to anisotropic ionospheres

The reader will have noticed that throughout the theoretical chapters of this monograph we have considered only one type of equation, namely

$$w'' + h^2 q w = 0. \tag{6.12}$$

W.K.B.J. solutions have been found in domains around the transition points defined to be the zeros of q. Connection formulae have been provided along the real axis, while reflection coefficients have been calculated by the method of phase integration.

Generalization may proceed in two directions. Either the equation may remain of the same form, but more complicated functions q may be considered possessing for example poles at various discrete points in the complex plane, or equations of higher order may be introduced. This was briefly discussed in the introductory chapter.

The equations governing radio propagation in an anisotropic ionosphere are essentially of the fourth order; these arise by including in the equation of motion of the free electrons a force due to the inter- action of their velocity with the earth's magnetic field. These fourth order equations may be expressed in first order form (similar to those used in section 2.5), using E_x, E_y, H_x, H_y as the four dependent variables (see Clemmow and Heading [25]). A matrix transformation then prepares the way for the production of four approximate W.K.B.J. solutions, which cease to be valid in the neighbourhood of transition points. These are points at which roots of a certain quartic equation become identical in value, usually in pairs. Such points are called *coupling points*, since two out of the four characteristic waves are associated in a metamorphosis that indissolubly links the solutions together at that point. It is possible to show by rearrangement (see

Heading [54]) that equations of the form (6.12) govern the meta-morphosis between the two characteristic waves, while the other two remain unaffected. The conclusion is made that the theory developed in this text may be applied immediately to examine the coupling between the two waves. Budden [24] in his text, *Radio Waves in the Ionosphere*, has examined this very fully and has considered a variety of problems using this method, though it is obvious as far as these generalizations are concerned that the mathematical theory is still far from complete.

The Relationship between Series and W.K.B.J. Solutions

A.1 The approximate value of an area

A graph $y = g(x)e^{f(x, h)} > 0$ is given for $x > 0$, such that when h is large e^f attains large values for a restricted range R of values of x, while for other values of x e^f is small; $g(x)$ varies but slowly throughout R. The full line in Fig. 35 represents the function considered. We seek an approximate value of the integral

$$I = \int_0^\infty ge^f dx,$$

the convergence of the integral being assumed.

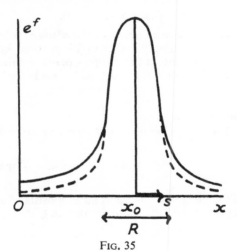

FIG. 35

We replace the hump (usually formed by a non-integrable function) by a similar hump (formed by an integrable function), the two curves differing only appreciably for values of x outside R. If the maximum value of $f(x)$ occurs at $x = x_0$, the first two non-vanishing terms of the Taylor-series expansion about the point $x = x_0$ are

$$f(x) = f(x_0) + \tfrac{1}{2}(x - x_0)^2 f''(x_0),$$

where $f''(x_0)$ is negative for a maximum. If $x - x_0 = s$, and if $g(x)$ is replaced by $g(x_0)$ throughout R, we have approximately

$$I \doteq g(x_0) \int\limits_{-\infty}^{+\infty} \exp\left[f(x_0) + \tfrac{1}{2}s^2 f''(x_0)\right] ds,$$

where the limits $\pm \infty$ are used in order to facilitate the integration. In the figure, the dotted line represents the curve under which we are calculating the area.

If $\tfrac{1}{2} f''(x_0) s^2 = -u^2$, we have

$$
\begin{aligned}
I &\doteq g(x_0) \exp\left[f(x_0)\right] \int\limits_{-\infty}^{\infty} \exp\left(-u^2\right) du / \sqrt{\left[-\tfrac{1}{2} f''(x_0)\right]} \\
&= \sqrt{\left[-2\pi / f''(x_0)\right]} g(x_0) \exp\left[f(x_0)\right],
\end{aligned}
\tag{A.1}
$$

using the standard infinite integral $\int \exp(-u^2) du = \sqrt{\pi}$.

This technique is a simplification of the method known as the *method of steepest descents* in the complex plane. Further correction terms may be obtained by using additional terms in the Taylor-series expansion.

A.2 The gamma function

If h is real and positive, $\Gamma(h)$ is defined by the value of the area

$$\Gamma(h) = \int\limits_0^\infty e^{-x} x^{h-1} dx \equiv \int\limits_0^\infty \exp\left[-x + (h-1)\log x\right] dx.$$

The graph of the integrand when h is large consists of a tall narrow peak. An integration by parts shows immediately that

$$\Gamma(h+1) = h\Gamma(h),$$

so

$$\Gamma(h) = \frac{1}{h} \int\limits_0^\infty \exp\left[-x + h\log x\right] dx.$$

Using the method developed in section A.1, we have

$$f(x) = -x + h \log x,$$
$$f'(x) = -1 + h/x,$$
$$f''(x) = -h/x^2.$$

$f'(x)$ vanishes when $x = h$, so $f(h) = -h + h \log h$ and $f''(h) = -1/h$; when h is large, the third derivative would be negligible. Equation (A.1) now gives

$$\Gamma(h) \rightleftharpoons h^{-1} \sqrt{(2\pi h)} \, e^{-h + h \log h} = \sqrt{(2\pi/h)} \, e^{-h} h^h. \qquad (A.2)$$

This approximation is known as *Stirling's formula*.

If h is complex, such that $-\pi < \arg h < \pi$, a more complete definition of $\Gamma(h)$ is required, but the same approximate formula is valid, *provided $\arg h$ lies within this restricted range.*

A.3 The Airy equation

The differential equation $w'' = zw$ possesses two simple power-series solutions, single-valued and convergent for all values of z. Term by term differentiation shows immediately that these two series are

$$w_1 = 1 + z^3/3! + 1.4z^6/6! + 1.4.7z^9/9! + \ldots,$$
$$w_2 = z + 2z^4/4! + 2.5z^7/7! + 2.5.8z^{10}/10! + \ldots = zv_2$$

say. The general solution is given by $w = A w_1 + B w_2$.

For large values of $|z|$, these series must be transformed into other forms in order to exhibit their true nature. When $z = h$ (real, large and positive) each term in w_1 is positive, and its approximate sum, following Stokes [106], may be found by replacing the discrete terms in the series by a continuous variable. The term involving h^{3n} is

$$\frac{1.4.7 \ldots (3n-2) h^{3n}}{(3n)!} = \frac{\frac{1}{3} \cdot \frac{4}{3} \cdot \frac{7}{3} \ldots (n - \frac{2}{3}) h^{3n} 3^n}{(3n)!} = \frac{\Gamma(n + \frac{1}{3}) h^{3n} 3^n}{\Gamma(\frac{1}{3}) \Gamma(3n+1)}.$$

We therefore consider the continuous graph

$$y = \frac{\Gamma(x + \frac{1}{3}) h^{3x} 3^x}{\Gamma(\frac{1}{3}) \Gamma(3x+1)},$$

the area under the graph for positive x being the approximate sum of the series w_1. The individual terms in w_1 (and hence the value of y)

increase to a maximum for some large value of n (or x); we shall therefore find the area under the graph by the method developed in section A.1.

When x is large, the use of formula (A.2) yields

$$y \rightleftharpoons \frac{\sqrt{[2\pi/(x+\tfrac{1}{3})]}\, e^{-x-1/3}\,(x+\tfrac{1}{3})^{x+1/3}\, h^{3x}\, 3^x}{\Gamma(\tfrac{1}{3})\sqrt{[2\pi/(3x+1)]}\, e^{-3x+1}\,(3x+1)^{3x+1}}$$

$$\rightleftharpoons \frac{x^{-2/3}}{\sqrt{3}\Gamma(\tfrac{1}{3})}\exp\left[-2x\log x - 2x\log 3 + 3x\log h + 2x\right],$$

where $x^{-2/3}$ is a slowly varying function contrasted to the exponent of the exponential. If

$$f = -2x\log x - 2x\log 3 + 3x\log h + 2x,$$

then

$$f' = -2\log x - 2 - 2\log 3 + 3\log h + 2,$$

$$f'' = -2/x.$$

Now f' vanishes when $x = \tfrac{1}{3}h^{3/2}$, at which point $f = \tfrac{2}{3}h^{3/2}$ and $f'' = -6h^{-3/2}$. Hence, using formula (A.1), we obtain

$$w_1 \rightleftharpoons [\sqrt{3}\Gamma(\tfrac{1}{3})]^{-1}(\tfrac{1}{3}h^{3/2})^{-2/3}\sqrt{(2\pi/6h^{-3/2})}\exp(\tfrac{2}{3}h^{3/2})$$

$$= [3^{-1/3}\sqrt{\pi}/\Gamma(\tfrac{1}{3})]\,h^{-1/4}\exp(\tfrac{2}{3}h^{3/2}), \tag{A.3}$$

a dominant expansion. This is one of the W.K.B.J. solutions of the Airy equation, but with a definite numerical coefficient.

Similarly, w_2 is replaced by

$$w_2(h) = h v_2 \rightleftharpoons h \int \frac{\Gamma(x+\tfrac{2}{3})}{\Gamma(\tfrac{2}{3})\,\Gamma(3x+2)}\, h^{3x}\, 3^x\, dx$$

where again h is real, large and positive. As before, the maximum of the integrand occurs when $x = \tfrac{1}{3}h^{3/2}$ yielding

$$w_2 \rightleftharpoons [3^{-2/3}\sqrt{\pi}/\Gamma(\tfrac{2}{3})]\,h^{-1/4}\exp(\tfrac{2}{3}h^{3/2}). \tag{A.4}$$

Two standard tabulated solutions [81] are taken to be

$$\mathrm{Ai}(z) = \tfrac{1}{2}\pi^{-1}3^{-1/6}\Gamma(\tfrac{1}{3})\,w_1 - \tfrac{1}{2}\pi^{-1}3^{1/6}\Gamma(\tfrac{2}{3})\,w_2, \tag{A.5}$$

$$\mathrm{Bi}(z) = \tfrac{1}{2}\pi^{-1}3^{1/3}\Gamma(\tfrac{1}{3})\,w_1 + \tfrac{1}{2}\pi^{-1}3^{2/3}\Gamma(\tfrac{2}{3})\,w_2, \tag{A.6}$$

for reasons now to be discussed.

When the expressions (A.3) and (A.4) are substituted into (A.5), we find that Ai(h) vanishes as far as these dominant terms are concerned, indicating that solution (A.5) must be subdominant along the positive real axis; this method does not however produce the subdominant expression.

Similarly, substitution into (A.6) yields after simplification

$$\text{Bi}(h) \doteq \pi^{-1/2} h^{-1/4} \exp\left(\tfrac{2}{3} h^{3/2}\right), \tag{A.7}$$

a simple dominant solution along the positive real axis.

When z is large, real and negative, the method by which summation is replaced by integration is not applicable, since the terms in the series are then alternately positive and negative. We therefore use an argument based upon the Stokes phenomenon discussed in Chapter III.

When $z = he^{2\pi i/3}$, all the terms in the series w_1 and v_2 are identical with the case when $z = h$, so their sums are still given by (A.3) and (A.4). In terms of z, the functional forms of the sums are obtained by replacing h by $ze^{-2\pi i/3}$, yielding

$$w_1 \doteq [3^{-1/3}\sqrt{\pi}/\Gamma(\tfrac{1}{3})] e^{\pi i/6} z^{-1/4} \exp\left(-\tfrac{2}{3} z^{3/2}\right), \tag{A.8}$$

and

$$w_2 = zv_2 \doteq [3^{-2/3}\sqrt{\pi}/\Gamma(\tfrac{2}{3})] e^{5\pi i/6} z^{-1/4} \exp\left(-\tfrac{2}{3} z^{3/2}\right), \tag{A.9}$$

yielding the dominant asymptotic expressions in the neighbourhood of the Stokes line at A in the figure.

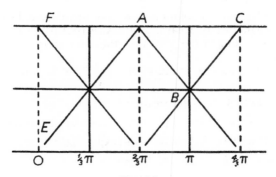

Fig. 36

When $z = he^{4\pi i/3}$, we use a similar argument. Replacing h by $ze^{-4\pi i/3}$, we obtain for the asymptotic expressions in the neighbourhood of the Stokes line at C:

$$w_1 \doteq [3^{-1/3}\sqrt{\pi}/\Gamma(\tfrac{1}{3})]e^{\pi i/3}z^{-1/4}\exp(\tfrac{2}{3}z^{3/2}), \tag{A.10}$$

$$w_2 = zv_2 \doteq [3^{-2/3}\sqrt{\pi}/\Gamma(\tfrac{2}{3})]e^{5\pi i/3}z^{-1/4}\exp(\tfrac{2}{3}z^{3/2}). \tag{A.11}$$

These two pairs of expressions continue to represent the series from A to B and from C to B, but at B, on the anti-Stokes line $\arg z = \pi$, the sum of the dominant expressions on both sides of the anti-Stokes line is required.

When $\arg z = \pi$,

$$\mathrm{Ai}(z) \doteq \tfrac{1}{2}\pi^{-1}3^{-1/6}\,\Gamma(\tfrac{1}{3})[\text{expression (A.8)} + \text{expression (A.10)}]$$

$$- \tfrac{1}{2}\pi^{-1}3^{1/6}\,\Gamma(\tfrac{2}{3})[\text{expression (A.9)} + \text{expression (A.11)}]$$

$$= \tfrac{1}{2}\pi^{-1/2}z^{-1/4}[i\exp(\tfrac{2}{3}z^{3/2}) + \exp(-\tfrac{2}{3}z^{3/2})]$$

upon simplification. Placing $z = he^{i\pi}$, we obtain

$$\mathrm{Ai}(-h) \doteq \pi^{-1/2}h^{-1/4}\cos(\tfrac{2}{3}h^{3/2} - \tfrac{1}{4}\pi). \tag{A.12}$$

Now for $\mathrm{Ai}(z)$, we have shown that there is no dominant solution at F where $\arg z = 0$; the subdominant solution at E is therefore obtained by using the same functional form as at A, namely

$$\mathrm{Ai}(z) \doteq \tfrac{1}{2}\pi^{-1}3^{-1/6}\,\Gamma(\tfrac{1}{3}) \times \text{expression (A.8)} - \tfrac{1}{2}\pi^{-1}3^{1/6}\,\Gamma(\tfrac{2}{3}) \times$$

$$\times \text{expression (A.9)}$$

$$= \tfrac{1}{2}\pi^{-1/2}z^{-1/4}\exp(-\tfrac{2}{3}z^{3/2}),$$

so
$$\mathrm{Ai}(h) \doteq \tfrac{1}{2}\pi^{-1/2}h^{-1/4}\exp(-\tfrac{2}{3}h^{3/2}), \tag{A.13}$$

a purely subdominant solution. We have therefore linked $\mathrm{Ai}(-h)$ (A.12) along the negative real axis with $\mathrm{Ai}(h)$ (A.13) along the positive real axis, providing at the same time the connection with the series solution (A.5).

Finally, $\mathrm{Bi}(z)$ may be found when $z = he^{i\pi}$; at B in the figure, we have

$$\mathrm{Bi}(z) \doteq \tfrac{1}{2}\pi^{-1}3^{1/3}\,\Gamma(\tfrac{1}{3})[\text{expression (A.8)} + \text{expression (A.10)}] +$$

$$+ \tfrac{1}{2}\pi^{-1}3^{2/3}\,\Gamma(\tfrac{2}{3})[\text{expression (A.9)} + \text{expression (A.11)}]$$

$$= \tfrac{1}{2}\pi^{-1/2}z^{-1/4}[\exp(\tfrac{2}{3}z^{3/2}) + i\exp(-\tfrac{2}{3}z^{3/2})],$$

$$\mathrm{Bi}(-h) \doteq \pi^{-1/2}h^{-1/4}\cos(\tfrac{2}{3}h^{3/2} + \tfrac{1}{4}\pi), \tag{A.14}$$

being $\tfrac{1}{2}\pi$ out of phase with $\mathrm{Ai}(-h)$.

A.4 The Weber equation

Similar methods may be used to calculate the Stokes constants for the equation possessing two transition points; in keeping with section 3.7 we shall write this equation in the form

$$w'' + (z^2 - a^2) w = 0,$$

where a^2 is real and positive, though the following argument is valid for complex values of a also. To simplify this equation, let

$$w = e^{-iz^2/2} u,$$

yielding $\qquad\qquad u'' - 2izu' - iu - a^2 u = 0.$

We may find two independent convergent power-series solutions of this equation by placing

$$u = z^c(1 + a_1 z^2 + a_2 z^4 + a_3 z^6 + \ldots),$$

inspection showing that all indices must differ from each other by an even integer. Substituting, and equating to zero the respective coefficients of powers of z, we obtain

$$c(c-1) = 0,$$

$$(c+2)(c+1)a_1 = 2ic + i + a^2,$$

$$(c+4)(c+3)a_2 = [2i(c+2) + i + a^2]a_1,$$

$$(c+6)(c+5)a_3 = [2i(c+4) + i + a^2]a_2.$$

When $c = 0$, we have

$$a_1 = (i + a_2)/2!,$$

$$a_2 = (5i + a^2)(i + a^2)/4!,$$

$$a_3 = (9i + a^2)(5i + a^2)(i + a^2)/6!,$$

and when $c = 1$, we have

$$a_1 = (3i + a^2)/3!,$$

$$a_2 = (7i + a^2)(3i + a^2)/5!,$$

$$a_3 = (11i + a^2)(7i + a^2)(3i + a^2)/7!.$$

The general power-series solution is therefore

$$w = A e^{-iz^2/2}\left[1 + \frac{i+a^2}{2!}z^2 + \frac{(5i+a^2)(i+a^2)}{4!}z^4 + \dots\right] +$$

$$+ B e^{-iz^2/2} z\left[1 + \frac{3i+a^2}{3!}z^2 + \frac{(7i+a^2)(3i+a^2)}{5!}z^4 + \dots\right]$$

$$\equiv A e^{-iz^2/2}v_1 + B e^{-iz^2/2}zv_2, \text{ say.} \tag{A.15}$$

Similarly, if we place $w = e^{iz^2/2}p$, we obtain

$$p'' + 2izp' + ip - a^2 p = 0,$$

possessing the general solution

$$w = C e^{iz^2/2}\left[1 - \frac{i-a^2}{2!}z^2 + \frac{(5i-a^2)(i-a^2)}{4!}z^4 - \dots\right] +$$

$$+ D e^{iz^2/2} z\left[1 - \frac{3i-a^2}{3!}z^2 + \frac{(7i-a^2)(3i-a^2)}{5!}z^4 - \dots\right]$$

$$\equiv C e^{iz^2/2}q_1 + D e^{iz^2/2}zq_2, \text{ say,} \tag{A.16}$$

each series now containing alternating signs.

If we place $z = 0$ in (A.15) and (A.16), we find $A = C$, while if we first differentiate with respect to z and place $z = 0$, we find $B = D$.

We shall now find the asymptotic expressions on each Stokes line for

$$w = e^{-iz^2/2}v_1 \equiv e^{iz^2/2}q_1.$$

When $z = e^{-i\pi/4}h$ or $e^{3i\pi/4}h$ (with a branch cut placed along the line $\arg z = \pi + \delta$ as in Fig. 11), we have

$$v_1 = 1 + \frac{i+a^2}{2!}(-i)h^2 + \frac{(5i+a^2)(i+a^2)}{4!}(-i)^2 h^4 + \dots,$$

the general term being

$$[(4n-3)i + a^2]\dots(i+a^2).(-i)^n h^{2n}/(2n)!$$

$$= (n - \tfrac{3}{4} - \tfrac{1}{4}ia^2)..(\tfrac{1}{4} - \tfrac{1}{4}ia^2).4^n h^{2n}/(2n)!$$

$$= \Gamma(n + \tfrac{1}{4} - \tfrac{1}{4}ia^2)4^n h^{2n}/\Gamma(\tfrac{1}{4} - \tfrac{1}{4}ia^2)\Gamma(2n+1). \tag{A.17}$$

Now formula (A.2) shows immediately that

$$\Gamma(n+\delta) = \Gamma(n)\, n^\delta$$

when n is large, so term (A.17) approximately equals

$$\Gamma(n)\, n^{1/4 - ia^2/4}\, 4^n\, h^{2n}/\Gamma(\tfrac{1}{4} - \tfrac{1}{4}ia^2)\, 2n\Gamma(2n),$$

which becomes when formula (A.2) is used

$$n^{-3/4 - ia^2/4}\, e^n\, n^{-n}\, h^{2n}/\sqrt{2}\, \Gamma(\tfrac{1}{4} - \tfrac{1}{4}ia^2).$$

We replace these discrete terms by a continuous function, but since a complex index occurs for n, the terms are not wholly real. But n raised to this constant index is a slowly varying function compared with the other variable terms, so we may replace n by x_0, a constant. Hence

$$v_1 \risingdotseq \frac{x_0^{-3/4 - ia^2/4}}{\sqrt{2}\, \Gamma(\tfrac{1}{4} - \tfrac{1}{4}ia^2)} \int \exp\left(x - x\log x + 2x\log h\right) dx,$$

so if

$$f = x - x\log x + 2x\log h,$$

$$f' = 1 - \log x - 1 + 2\log h,$$

$$f'' = -1/x,$$

then $f' = 0$ when $x_0 = h^2$ and $f(x_0) = h^2$, $f''(x_0) = -1/h^2$.

Formula (A.1) then yields

$$v_1 \risingdotseq h^{-3/2 - ia^2/2}\sqrt{(2\pi h^2)}\, e^{h^2}/\sqrt{2}\, \Gamma(\tfrac{1}{4} - \tfrac{1}{4}ia^2),$$

so

$$w = e^{-iz^2/2}v_1 \risingdotseq \sqrt{\pi}\, h^{-1/2 - ia^2/2}\, e^{h^2/2}/\Gamma(\tfrac{1}{4} - \tfrac{1}{4}ia^2), \qquad (A.18)$$

when $z = e^{-i\pi/4}\, h$ or $e^{3i\pi/4}\, h$.

When $z = e^{i\pi/4}h$ or $e^{-3i\pi/4}h$, we must use the series q_1 instead of v_1, since v_1 would contain alternating signs, but q_1 now has similar signs throughout. An inspection of the series q_1 shows that its value is merely the conjugate of v_1 for these values of z, so

$$w \risingdotseq \sqrt{\pi}\, h^{-1/2 + ia^2/2}\, e^{h^2/2}/\Gamma(\tfrac{1}{4} + \tfrac{1}{4}ia^2). \qquad (A.19)$$

The Stokes constants may now be calculated. Replacing h by its respective value in terms of z, we obtain

$$w \risingdotseq \sqrt{\pi}\, (e^{3i\pi/4})^{-1/2 + ia^2/2}\, z^{-1/2 + ia^2/2}\, e^{-iz^2/2}/\Gamma(\tfrac{1}{4} + \tfrac{1}{4}ia^2),$$

$$\arg z = -\tfrac{3}{4}\pi,$$

$$w \doteq \sqrt{\pi}\, (e^{i\pi/4})^{-1/2 - ia^2/2} z^{-1/2 - ia^2/2} e^{iz^2/2}/\Gamma(\tfrac{1}{4} - \tfrac{1}{4}ia^2),$$
$$\arg z = -\tfrac{1}{4}\pi,$$

$$w \doteq \sqrt{\pi}\, (e^{-i\pi/4})^{-1/2 + ia^2/2} z^{-1/2 + ia^2/2} e^{-iz^2/2}/\Gamma(\tfrac{1}{4} + \tfrac{1}{4}ia^2),$$
$$\arg z = \tfrac{1}{4}\pi,$$

$$w \doteq \sqrt{\pi}\, (e^{-3i\pi/4})^{-1/2 - ia^2/2} z^{-1/2 - ia^2/2} e^{iz^2/2}/\Gamma(\tfrac{1}{4} - \tfrac{1}{4}ia^2),$$
$$\arg z = \tfrac{3}{4}\pi.$$

If U is the Stokes constant when $\arg z = \tfrac{1}{4}\pi$, we have

dominant coefficient when $\arg z = \tfrac{3}{4}\pi$

$\qquad =$ dominant coefficient when $\arg z = -\tfrac{1}{4}\pi$

$\qquad\qquad + U \times$ dominant coefficient when $\arg z = \tfrac{1}{4}\pi$.

Hence

$$U = \frac{\Gamma(\tfrac{1}{4} + \tfrac{1}{4}ia^2)}{\Gamma(\tfrac{1}{4} - \tfrac{1}{4}ia^2)} \cdot \frac{e^{3i\pi(1/2 + ia^2/2)/4} - e^{-i\pi(1/2 + ia^2/2)/4}}{e^{i\pi(1/2 - ia^2/2)/4}}$$

$$= \frac{\Gamma(\tfrac{1}{4} + \tfrac{1}{4}ia^2)}{\Gamma(\tfrac{1}{4} - \tfrac{1}{4}ia^2)} e^{-\pi a^2/4} \cdot 2i \sin\left(\tfrac{1}{4}\pi + \tfrac{1}{4}\pi ia^2\right)$$

$$= \frac{2\pi i e^{-\pi a^2/4}}{\Gamma(\tfrac{1}{4} - \tfrac{1}{4}ia^2)\,\Gamma(\tfrac{3}{4} - \tfrac{1}{4}ia^2)},$$

using the standard formula $\Gamma(z)\Gamma(1 - z) = \pi/\sin\pi z$. Finally, the use of the duplication formula

$$\Gamma(z)\,\Gamma(\tfrac{1}{2} + z) = \sqrt{\pi}\, \Gamma(2z)/2^{2z - 1}$$

yields $\qquad U = i\sqrt{(2\pi)}\, 2^{-ia^2/2} e^{-\pi a^2/4}/\Gamma(\tfrac{1}{2} - \tfrac{1}{2}ia^2),$

as stated in (3.17). Similarly, the other Stokes constants may be found, yielding the results (3.18).

A.5. Tables of the Airy integral

The author acknowledges with grateful thanks the kind permission granted by the Chairman of the Mathematical Tables Committee of the Royal Society and by the compiler Dr. J. C. P. Miller to reproduce the following small part of the British Association tables of the Airy Integral [81]. The table is included to provide the reader with a feeling of the functions $\mathrm{Ai}(x)$, $\mathrm{Ai}'(x)$, $\mathrm{Bi}(x)$ and $\mathrm{Bi}'(x)$ in the main region of metamorphosis along the real axis. Three decimal places are given, and x lies in the range $-5 \cdot 0(0 \cdot 2) - 1 \cdot 0(0 \cdot 1) + 1 \cdot 0(0 \cdot 2) + 2 \cdot 0$. Further tables have recently been compiled by Smirnov [126].

x	Ai(x)	Ai$'(x)$	Bi(x)	Bi$'(x)$
-5·0	+0·351	+0·327	-0·138	+0·778
-4·8	+0·380	-0·037	+0·026	+0·835
-4·6	+0·337	-0·380	+0·185	+0·735
-4·4	+0·234	-0·641	+0·311	+0·509
-4·2	+0·089	-0·782	+0·383	+0·206
-4·0	-0·070	-0·791	+0·392	-0·117
-3·8	-0·219	-0·677	+0·339	-0·406
-3·6	-0·335	-0·470	+0·235	-0·621
-3·4	-0·403	-0·208	+0·097	-0·739
-3·2	-0·417	+0·065	-0·054	-0·754
-3·0	-0·379	+0·315	-0·198	-0·676
-2·8	-0·295	+0·512	-0·319	-0·524
-2·6	-0·178	+0·642	-0·405	-0·327
-2·4	-0·043	+0·698	-0·449	-0·112
-2·2	+0·096	+0·686	-0·450	+0·096
-2·0	+0·227	+0·618	-0·412	+0·279
-1·8	+0·341	+0·510	-0·341	+0·423
-1·6	+0·430	+0·379	-0·246	+0·524
-1·4	+0·492	+0·240	-0·135	+0·582
-1·2	+0·526	+0·107	-0·016	+0·602
-1·0	+0·536	-0·010	+0·104	+0·592
-0·9	+0·532	-0·061	+0·163	+0·580
-0·8	+0·524	-0·106	+0·220	+0·564
-0·7	+0·511	-0·145	+0·275	+0·545
-0·6	+0·495	-0·177	+0·329	+0·525
-0·5	+0·476	-0·204	+0·380	+0·506
-0·4	+0·454	-0·225	+0·430	+0·488
-0·3	+0·431	-0·241	+0·478	+0·472
-0·2	+0·406	-0·251	+0·524	+0·459
-0·1	+0·381	-0·257	+0·570	+0·451
0·0	+0·355	-0·259	+0·615	+0·448
0·1	+0·329	-0·257	+0·660	+0·452
0·2	+0·304	-0·252	+0·705	+0·462
0·3	+0·279	-0·245	+0·752	+0·480
0·4	+0·255	-0·236	+0·802	+0·507
0·5	+0·232	-0·225	+0·854	+0·545
0·6	+0·210	-0·213	+0·911	+0·593
0·7	+0·189	-0·200	+0·973	+0·654
0·8	+0·170	-0·186	+1·042	+0·730
0·9	+0·152	-0·173	+1·120	+0·822
1·0	+0·135	-0·159	+1·207	+0·932
1·2	+0·106	-0·133	+1·421	+1·221
1·4	+0·082	-0·108	+1·704	+1·627
1·6	+0·063	-0·087	+2·082	+2·193
1·8	+0·047	-0·069	+2·596	+2·986
2·0	+0·035	-0·053	+3·298	+4·101

Bibliography

1. ABBOTT, M. R. (1956). *Proc. Camb. Phil. Soc.*, **52**, 344–362.
2. ARNOT, F. L., and BAINES, G. O. (1934). *Proc. Roy. Soc.*, A, **146**, 651–662.
3. ATKINSON, F. V. (1960). *J. Math. Anal. & Applns.*, **1**, 255–276.
4. AULUCK, F. C., and KOTHARI, D. S. (1945). *Proc. Camb. Phil. Soc.*, **41**, 175–179.
5. BAILEY, V. A. (1954). *Phys. Rev.*, **96**, 865–868.
6. BAILEY, V. A. (1954). *Scientific Report No. 67, Ionospheric Research Laboratory, Pennsylvania State University.*
7. BELL, R. P. (1944). *Phil. Mag.*, **35**, (7), 582–588.
8. BELLMAN, R., and KALABA, R. (1958). *Proc. Nat. Acad. Sci.*, **44**, 317–319.
9. BELLMAN, R., and KALABA, R. (1959). *J. Math. Mech.*, **8**, 683–704.
10. BENNEY, D. J. (1961). *J. Fluid Mech.*, **10**, 209–236.
11. BIRKHOFF, G. D. (1908). *Trans. Amer. Math. Soc.*, **9**, 219–231.
12. BIRKHOFF, G. D. (1933). *Bull. Amer. Math. Soc.*, **39**, 681–700.
13. BOHM, D. (1951). *Quantum Theory.* Prentice-Hall Inc., New York.
14. BOOKER, H. G. (1938). *Phil. Trans. Roy. Soc.* A, **237**, 411–451.
15. BOOKER, H. G., and WALKINSHAW, W. (1946). *Meteorological Factors in Radio-wave Propagation*, 80–127. The Physical Society.
16. BREMMER, H. (1949). *Physica*, **15**, 593–608.
17. BREMMER, H. (1949). *Philips Res. Rep.*, **4**, 1–19.
18. BREMMER, H. (1949). *Philips Res. Rep.*, **4**, 189–205.
19. BREMMER, H. (1949). *Terrestrial Radio Waves.* Elsevier Pub. Co. Inc.
20. BREMMER, H. (1951). *Comm. Pure & Appl. Maths.*, **4**, 105–115.
21. BRILLOUIN, L. (1926). *C. R. Acad. Sci. Paris*, **183**, 24–26.
22. BUDDEN, K. G. (1952). *Proc. Roy. Soc.* A, **215**, 215–233.
23. BUDDEN, K. G. (1955). *Proc. Roy. Soc.* A, **227**, 516–537.
24. BUDDEN, K. G. (1961). *Radio Waves in the Ionosphere*, Cambridge.
25. CLEMMOW, P. C., and HEADING, J. (1954), *Proc. Camb. Phil. Soc.*, **50**, 319–333.

152

26. DINGLE, R. B. (1957). *Proc. Roy. Soc.* A, **244**, 456–475.
27. DINGLE, R. B. (1958). *Proc. Roy. Soc.* A, **249**, 270–283.
28. DUNHAM, J. L. (1932). *Phys. Rev.*, **41**, 721–731.
29. ECKART, C. (1960). *Hydrodynamics of Oceans and Atmospheres.* Pergamon Press.
30. ECKERSLEY, T. L. (1931). *Proc. Roy. Soc.* A, **132**, 53.
31. ECKERSLEY, T. L. (1932). *Proc. Roy. Soc.* A, **136**, 499.
32. ECKERSLEY, T. L. (1950). *Proc. Phys. Soc.* B, **63**, 49.
33. ECKERSLEY, T. L., and MILLINGTON G. (1938). *Phil. Trans. Roy. Soc.* A, **237**, 273.
34. FÖRSTERLING, K. (1942). *Hochfrequenztech. u. Elektroakust.*, **59**, 10.
35. FOWLER, R. H. *et al.* (1920). *Phil. Trans. Roy. Soc.* A, **221**, 337–339.
36. FRIEDMAN, B. (1951). *Comm. Pure & Appl. Maths.*, **4**, 317–350.
37. FURRY, W. H. (1947). *Phys. Rev.*, **71**, 360–371.
38. GAMOW, G. (1928). *Zeit. f. Phys.*, **51**, 204–212.
39. GAMOW, G. (1937). *Atomic Nuclei and Nuclear Transformations.* Oxford.
40. GAMOW, G., and CRITCHFIELD, C. L. (1949). *Theory of Atomic Nucleus and Nuclear Energy Sources.* Oxford.
41. GANS, R. (1915). *Ann. Phys. Lpz.*, **47**, (4), 709–736.
42. GIBBONS, J. J., and NERTNEY, R. J. (1951). *J. Geophys. Res.*, **56**, 355–371.
43. GIBBONS, J. J., and NERTNEY, R. J. (1952). *J. Geophys. Res.*, **57**, 323–338.
44. GIBBONS, J. J., and SCHRAG, R. L. (1952). *J. Appl. Phys.*, **23**, 1139–1142.
45. GOL'DMAN, I. I. *et al.* (1960). *Problems in Quantum Mechanics.* Infosearch, London.
46. GOLDSTEIN, S. (1928). *Proc. Lond. Math. Soc.*, **28**, 81–90.
47. GOLDSTEIN, S. (1931). *Proc. Lond. Math. Soc.*, **33**, 246–252.
48. GREEN, G. (1837). *Camb. Phil. Trans.*, **6**, 457–462.
49. HARTREE, D. R. (1931). *Proc. Roy. Soc.* A, **131**, 428.
50. HEADING, J. (1953). *Theoretical Ionospheric Radio Propagation.* Thesis, Cambridge.
51. HEADING, J. (1957). *Proc. Camb. Phil. Soc.*, **53**, 399–418.
52. HEADING, J. (1957). *Proc. Camb. Phil. Soc.*, **53**, 419–441.
53. HEADING, J. (1960). *Proc. Camb. Phil. Soc.*, **56**, 329–341.
54. HEADING, J. (1961). *J. Res. Nat. Bur. Standards* D, **65**, 595–616.
55. HEADING, J. (1962). To be published in *J. Lond. Math. Soc.*

56. HEADING, J. (1962). To be published in *Quart. J. Mech. & Appl. Maths.*
57. HEADING, J. (1962). To be published in *Math. Gaz.*
58. HEISENBERG, W. (1924). *Ann. Phys.*, **74**, 577.
59. HINES, C. O. (1953). *Quart. Appl. Maths.*, **11**, 9–31.
60. HORN, J. (1899). *Math. Ann.*, **52**, 271–292.
61. IAMI, I. (1948). *Phys. Rev.*, **74**, 113.
62. IAMI, I. (1958). *Quart. Appl. Maths.*, **16**, 33–45.
63. JEFFREYS, B. (1942). *Proc. Camb. Phil. Soc.*, **38**, 401–405.
64. JEFFREYS, H. (1923). *Proc. Lond. Math. Soc.*, **23**, (2), 428–436.
65. JEFFREYS, H. (1942). *Phil. Mag.*, **33**, (7), 451–456.
66. JEFFREYS, H. (1953). *Proc. Camb. Phil. Soc.*, **49**, 601–611.
67. JEFFREYS, H. (1956). *Proc. Camb. Phil. Soc.*, **52**, 61–66.
68. JEFFREYS, H., and JEFFREYS, B. (1956). *Methods of Mathematical Physics.* Cambridge.
69. JENSSEN, O. (1960). *J. Maths. & Phys.*, **39**, 1–17.
70. KEMBLE, E. C. (1937). *The Fundamental Principles of Quantum Mechanics.* McGraw-Hill.
71. KERR, O. E. (1951). *Propagation of Short Radio Waves.* Rad. Lab. Series.
72. KRAMERS, H. A. (1926). *Zeit. f. Phys.*, **39**, 828–840.
73. LANDAU, L. D. and LIFSHITZ, E. M. (1958). *Quantum Mechanics.* Pergamon Press.
74. LANGER, R. E. (1934). *Bull. Amer. Math. Soc.* (2), **40**, 545–582.
75. LANGER, R. E. (1937). *Phys. Rev.*, **51**, 669–676.
76. LANGER, R. E. (1950). *Comm. Pure & Appl. Maths.*, **3**, 427–438.
77. LANGER, R. E. (1957). *Trans. Amer. Math. Soc.*, **84**, 144–191.
78. LIN, C. C. (1945). *Quart. Appl. Maths.*, **3**, 117–142.
79. LIOUVILLE, J. (1837). *J. Math. Pures Appl.*, **2**, 16–35.
80. MEKSYN, D. (1946). *Proc. Roy. Soc.* A, **187**, 492–504.
81. MILLER, J. C. P. (1946). British Association Tables, Part-volume B, *The Airy Integral.*
82. MORIGUCHI, H. (1960). *Prog. Theor. Phys.*, **23**, 750–752.
83. MORSE, P. M., and FESHBACH, H. (1953). *Methods of Theoretical Physics.* McGraw-Hill.
84. MOTT, N. F. (1928). *Proc. Camb. Phil. Soc.*, **24**, 76–79.
85. MOTT, N. F., and MASSEY, H. S. W. (1933). *The Theory of Atomic Collisions.* Oxford.
86. MOTT, N. F., and SNEDDON, I. N. (1948). *Wave Mechanics and its Applications.* Oxford.
87. NERTNEY, R. J. (1952). *J. Geophys. Res.*, **57**, 423–425.

88. OLVER, F. W. J. (1959). *J. Soc. Indust. Appl. Maths.* **7**, 306–310.
89. OLVER, F. W. J. (1959). *J. Res. Nat. Bur. Standards* B, **63**, 131–169.
90. OLVER, F. W. J. (1961). *Proc. Camb. Phil. Soc.*, **57**, 790–810.
91. PAULI, W. (1958). *Handbuch der Physik*, **5**, (1), 94.
92. PFISTER, W. (1953). *J. Geophys. Res.*, **58**, 29–40.
93. RABENSTEIN, A. L. (1958). *Arch. Rat. Mech. & Anal.*, **1**, 418–435.
94. RAYLEIGH, LORD (1912). *Proc. Roy. Soc.* A, **86**, 207–226.
95. RYDBECK, O. E. H. (1948). *Trans. Chalmers Univ.*, *Sweden*, No. 74.
96. RYDBECK, O. E. H. (1951) *Comm. Pure & Appl. Maths.*, **4**, 129–160.
97. SCHELKUNOFF, S. A. (1951). *Comm. Pure & Appl. Maths.*, **4**, 117–128.
98. SCHLESINGER, L. (1907). *Math. Ann.*, **63**, 277–300.
99. SECKLER, B. D., and KELLER, J. B. (1959). *J. Acoustical Soc. Amer.*, **31**, 206–216.
100. SLATER, J. L. (1951). *Quantum Theory of Matter*. McGraw-Hill.
101. SYMTH, J. B. (1952). *J. Geophys. Res.*, **57**, 423.
102. SYMTH, J. B. (1952). *J. Geophys. Res.*, **57**, 425–426.
103. STOKES, G. G. (1907). *Sir George Gabriel Stokes, Memoirs and Scientific Correspondence*, Vol. 1, p. 62. Cambridge.
104. STOKES, G. G. (1857). *Trans. Camb. Phil. Soc.*, **10**, 106–128.
105. STOKES, G. G. (1871). *Trans. Camb. Phil. Soc.*, **11**, 412–425.
106. STOKES, G. G. (1889). *Proc. Camb. Phil. Soc.*, **6**, 362–366.
107. TAMARKIN, J. (1928). *Math. Zeit.*, **27**, 1–54.
108. *Technology* (1961), **5**, 193.
109. TITCHMARSH, E. C. (1946). *Eigenfunction Expansions*, Vol. 1. Oxford.
110. TURRITTIN, H. L. (1936). *Amer. J. Maths.*, **58**, 364–376.
111. WASOW, W. (1950). *Ann. of Maths.*, **52**, 350–361.
112. WASOW, W. (1953). *Ann. of Maths.*, **58**, 222–252.
113. WASOW, W. (1960). *J. Maths. & Phys.*, **38**, 257–278.
114. WATSON, G. N. (1944). *Bessel Functions*. Cambridge.
115. WENTZEL, G. (1926). *Zeit. f. Phys.*, **38**, 518.
116. WHITTAKER, E. T., and WATSON, G. N. (1927). *A Course of Modern Analysis*. Cambridge.
117. YOUNG, L. A., and UHLENBECK, G. E. (1930). *Phys. Rev.*, **36**, 1154–1167.
118. ZWAAN, A. (1929). *Intensitäten im Ca-Funkenspektrum*, Thesis. Utrecht.

Additional References

119. SINGH, K. K., and BAIJAL, J. S. (1955). *Prog. Theor. Phys.*, **14**, 214.

120. HULL, T. E., and JULIUS, R. S. (1956). *Can. J. Phys.*, **34**, 914–919.

121. SINGH, K. K. (1959). *Proc. Nat. Sci. India* A, **25**, 295.

122. TAKAHASHI, T. (1955). *Bull. Earthquake Res. Inst.*, **33**, 287–296.

123. TAKAHASHI, T. (1957). *Bull. Earthquake Res. Inst.*, **35**, 297–308.

124. BREKHOVSKIKH, L. M. (1960). *Waves in Layered Media.* Academic Press Inc., New York.

125. BUDDEN, K. G. (1961). *Proc. Roy. Soc.* A, **263**, 552–566.

126. SMIRNOV, A. D. (1960). *Tables of Airy Integrals and Special Confluent Hypergeometric Functions.* Pergamon Press.

127. INCE, E. L. (1956). *Ordinary Differential Equations.* Dover Publications.

128. BUDDEN, K. G. (1961). *The Wave-guide Mode Theory of Wave Propagation.* Prentice-Hall.

129. GINZBURG, V. L. (1961). *Propagation of Electromagnetic Waves in Plasma.* North-Holland Publishing Company.

130. JEFFREYS, H. (1962). *Asymptotic Approximations.* Oxford.

Index

Abbott, M. R., 19
Acoustical diffraction, 19
Aerial, 18, 21
Airy equation, 11, 23, 49
  standard solutions of, 59–61,
    143–146
Airy integral, 4, 8, 20, 29, 45
  tables of, 151
Anisotropic ionosphere, 22, 23, 134,
    139
Anti-Stokes lines, 8, 13, 32, 69
  rules for crossing, 54
Approximations, 2
  additional, 21
Area, 141
Arnot, F. L., 18
Asymptotic series, 5, 9, 59
  uniformity, 8, 33
Atkinson, F. V., 20
Auluck, F. C., 15, 127
Average value, 89, 127

Bailey, V. A., 3, 22
Baines, G. O., 18
Bell, R. P., 15
Bellman, R., 20, 134
Benney, D. J., 21
Bessel's equation, 3, 5, 44, 61–63,
    84, 90
  functions, 7, 45, 84
Birkhoff, G. D., 9, 14
Booker, H. G., 19, 23, 134
Born approximation, 18
Bounded harmonic oscillator, 15,
    123–127
Branch cuts, 40
  rules for crossing, 53, 71, 90
Brekhovskikh, L. M., 19, 156
Bremmer, H., 19, 20, 23, 131
  interpretation, 131–134

Brillouin, L., 3
Budden, K. G., 5, 8, 21, 23, 38, 75,
    134, 140, 156

Carlini, 3
Cauchy, 47
  theorem, 86
Characteristic roots, 37
  waves, 38
Classical limit, 116
Clemmow, P. C., 139
Comparison equations, 27
Complex barrier, 101–102
  phase-integral, 87
  plane, 52
Confluent hypergeometric function,
    6
Connection formulae, 9, 14, 19,
    77–81, 89, 119
Conservation of energy flow, 75, 87
Contour integration, 52
Converging factors, 6
Coupled equations, 22, 38
Coupling points, 38, 139
  terms, 38

Diffraction, 18
Dingle, R. B., 6
Direction of propagation, 73–75
Domains of validity, 21, 33, 35
Dominancy changing factor, 96
Dominant solutions, 9, 31, 40, 78
Dunham, J. L., 14

Eckart, C., 14, 130, 134
Eckersley, T. L., 13, 18, 23, 85
Eigenvalues, 21, 95, 104, 115, 123,
    131
Electromagnetic waves, 134
Electronic computer, 23

Electron scattering, 19
Energy flow, 12, 19, 75, 87
Error bounds, 7
    terms, 32, 34, 50, 51
Evanescent wave, 42
Existence theorems, 3

Försterling, K., 22
Fourth-order equations, 23, 139
Fowler, R. H., 7
Freehafer, 14
Friedman, B., 19
Furry, W. H., 13, 15, 71, 106

Gamma function, 67, 142
Gamow, G., 16
Gand, R., 6, 10
Gibbons, J. J., 22, 23
Ginzberg, V. L., 156
Gol'dman, I. I., 19, 119
Goldstein, S., 13, 89
Green, G., 3
Group velocity, 138

Hamilton-Jacobi equation, 136
Hankel functions, 45, 84
Harmonic oscillator, 14, 95, 124
    approximate, 102–105
    bounded, 15, 123–127
Hartree, D. R., 16, 17
Heading, J., 6, 8, 21, 22, 36, 59, 139,
    140
    rule, 75
Heat transfer, 22
Heisenberg, W., 24
Hermite polynomials, 104, 109
Higher approximations, 20
Hines, C. O., 2, 22
Historical survey, 1–24
Horn, J., 4, 9
Hydrodynamics, 24
Hydrogen atom, 15, 127–131
Hyperbolic functions, 48

Iami, I., 21
Ince, E. L., 3, 156
Incident wave, 81
Integral equation, 29, 50
    solution of, 31
Ionospheric propagation, 16, 17, 18,
    72, 134–140
Iterative process, 22

Jeffreys, B., 12, 16, 17, 89, 95
Jeffreys, H., 3, 4, 7, 8, 10, 11, 16, 20,
    21, 36, 97, 156
    connection formulae, 77–81, 89,
    119
    form, 39
    recommendation, 39
Jenssen, O., 22

Kemble, E. C., 13, 16
Kerr, O. E., 13
Kothari, D. S., 15, 127
Kramers, H. A., 3, 13

Landau, L. D. and Lifshitz, E. M.,
    16, 128, 131
Langer, R. E., 9, 11, 13, 16, 20, 21,
    24, 128
Light, 7, 10
Limiting polarization, 23
Lin, C. C., 24
Liouville, J., 3
Loss-free medium, 76

Mathieu's equation, 7
Matrix notation, 9, 36, 121, 139
Maxwell's equations, 134
Meksyn, D., 24
Membrane, 6
Metamorphosis, 9, 83, 140
Microwave propagation, 21
Miller, J. C. P., 47
Millington, G., 18
Moriguchi, H., 22
Mott, N. F., 11
Multiple reflection, 132

Necessary conditions, 28
Nertney, R. J., 22
Neutral solution, 40
Non-linear theory, 21
Normalization, 15, 105–110
  unsatisfactory, 105
Notation, 40

Olver, F. W. J., 7, 8, 9, 21
Overdense, 72, 95–97

Parabolic cylinder functions, 7, 101, 147–150
Patching methods, 10, 20
Pauli, W., 105
Perfect reflector, 12
Periodic function, example, 118–123
Phase-integral methods, 73
Phase-reference level, 81
Planck's hypothesis, 18, 118
Poles, 22
Potential barrier, 12, 16, 94
  overdense, 95–99
  underdense, 99–101
Potential well, 14, 94
Power series, 62, 143, 147
Poynting vector, 75, 138
Progressive waves, 43

Rabenstein, A. L., 24
Rayleight, Lord, 3, 6, 10
Ray methods, 17, 23, 134–139
  path, 137
Reflection coefficient, 21, 22, 81–83, 92, 97, 101, 102, 111, 113
  example, 83
  phase integration, 85
  points, 38
Refractive index, 10, 40
Ricatti equation, 25–27, 136
Riemann-Cauchy equations, 74
Riemann sheets, 23
Rydbeck, O. E. H., 23

Schelkunoff, S. A., 2
Schlesinger, L., 9
Schrödinger's equation, 13, 14, 18, 116
Seckler, B. D., 19
Seismology, 21
Semi-classical approximation, 19
Simultaneous equations, 37, 139
Slater, J. L., 14
Smirnov, A. D., 150, 156
Smyth, J. B., 2, 22
Standing waves, 12
Steepest descents, 142
Stirling's formula, 143
Stokes, G. G., 4, 104
  constants, 6, 55, 57, 59, 63, 67, 71, 87, 90, 150
  lines, 5, 13, 55, 70
    rules for crossing, 55
  phenomenon, 4, 13, 48, 56
    diagram showing, 60, 65, 145
Stratified medium, 7, 19, 135
Subdominant solutions, 9, 31, 40, 78

Takahashi, T., 21
Tamarkin, J., 9
Thermocline, 15
Thermosphere, 15, 130
Tidal waves, 4
Titchmarsh, E. C., 15
Toroidal shells, 22
Transition point, 2, 32, 38
  approximation valid at, 20
  one, 69–93
  order greater than one, 13, 89, 110–115
  two, 64–68, 94–115
Transmission coefficients, 92, 97, 101, 102, 113
Triple splitting, 23
Tropospheric refraction, 19
Turrittin, H. L., 9

Uhlenbeck, G. E., 16
Underdense, 72
Uniform approximations, 8, 33

Variation of parameters, 13, 22, 30

Walkinshaw, W., 19
Wasow, W., 19, 20, 24
Watson, G. N., 3, 18
Wave packet, 23, 136

Waves, 72
    direction of propagation, 73–75
Weber's equations, 64, 147–150
Wentzel, G., 3, 13
W.K.B.J. approximations, 6, 25, 69
    examples, 41–47
Wronksian, 26, 20

Young, L. A., 16

Zwaan, A., 13

A CATALOG OF SELECTED
DOVER BOOKS
IN SCIENCE AND MATHEMATICS

Astronomy

CHARIOTS FOR APOLLO: The NASA History of Manned Lunar Spacecraft to 1969, Courtney G. Brooks, James M. Grimwood, and Loyd S. Swenson, Jr. This illustrated history by a trio of experts is the definitive reference on the Apollo spacecraft and lunar modules. It traces the vehicles' design, development, and operation in space. More than 100 photographs and illustrations. 576pp. 6 3/4 x 9 1/4.   0-486-46756-2

EXPLORING THE MOON THROUGH BINOCULARS AND SMALL TELESCOPES, Ernest H. Cherrington, Jr. Informative, profusely illustrated guide to locating and identifying craters, rills, seas, mountains, other lunar features. Newly revised and updated with special section of new photos. Over 100 photos and diagrams. 240pp. 8 1/4 x 11.   0-486-24491-1

WHERE NO MAN HAS GONE BEFORE: A History of NASA's Apollo Lunar Expeditions, William David Compton. Introduction by Paul Dickson. This official NASA history traces behind-the-scenes conflicts and cooperation between scientists and engineers. The first half concerns preparations for the Moon landings, and the second half documents the flights that followed Apollo 11. 1989 edition. 432pp. 7 x 10.
0-486-47888-2

APOLLO EXPEDITIONS TO THE MOON: The NASA History, Edited by Edgar M. Cortright. Official NASA publication marks the 40th anniversary of the first lunar landing and features essays by project participants recalling engineering and administrative challenges. Accessible, jargon-free accounts, highlighted by numerous illustrations. 336pp. 8 3/8 x 10 7/8.   0-486-47175-6

ON MARS: Exploration of the Red Planet, 1958-1978--The NASA History, Edward Clinton Ezell and Linda Neuman Ezell. NASA's official history chronicles the start of our explorations of our planetary neighbor. It recounts cooperation among government, industry, and academia, and it features dozens of photos from Viking cameras. 560pp. 6 3/4 x 9 1/4.   0-486-46757-0

ARISTARCHUS OF SAMOS: The Ancient Copernicus, Sir Thomas Heath. Heath's history of astronomy ranges from Homer and Hesiod to Aristarchus and includes quotes from numerous thinkers, compilers, and scholasticists from Thales and Anaximander through Pythagoras, Plato, Aristotle, and Heraclides. 34 figures. 448pp. 5 3/8 x 8 1/2.
0-486-43886-4

AN INTRODUCTION TO CELESTIAL MECHANICS, Forest Ray Moulton. Classic text still unsurpassed in presentation of fundamental principles. Covers rectilinear motion, central forces, problems of two and three bodies, much more. Includes over 200 problems, some with answers. 437pp. 5 3/8 x 8 1/2.   0-486-64687-4

BEYOND THE ATMOSPHERE: Early Years of Space Science, Homer E. Newell. This exciting survey is the work of a top NASA administrator who chronicles technological advances, the relationship of space science to general science, and the space program's social, political, and economic contexts. 528pp. 6 3/4 x 9 1/4.
0-486-47464-X

STAR LORE: Myths, Legends, and Facts, William Tyler Olcott. Captivating retellings of the origins and histories of ancient star groups include Pegasus, Ursa Major, Pleiades, signs of the zodiac, and other constellations. "Classic." – *Sky & Telescope*. 58 illustrations. 544pp. 5 3/8 x 8 1/2.   0-486-43581-4

A COMPLETE MANUAL OF AMATEUR ASTRONOMY: Tools and Techniques for Astronomical Observations, P. Clay Sherrod with Thomas L. Koed. Concise, highly readable book discusses the selection, set-up, and maintenance of a telescope; amateur studies of the sun; lunar topography and occultations; and more. 124 figures. 26 halftones. 37 tables. 335pp. 6 1/2 x 9 1/4.   0-486-42820-6

Chemistry

MOLECULAR COLLISION THEORY, M. S. Child. This high-level monograph offers an analytical treatment of classical scattering by a central force, quantum scattering by a central force, elastic scattering phase shifts, and semi-classical elastic scattering. 1974 edition. 310pp. 5 3/8 x 8 1/2.   0-486-69437-2

HANDBOOK OF COMPUTATIONAL QUANTUM CHEMISTRY, David B. Cook. This comprehensive text provides upper-level undergraduates and graduate students with an accessible introduction to the implementation of quantum ideas in molecular modeling, exploring practical applications alongside theoretical explanations. 1998 edition. 832pp. 5 3/8 x 8 1/2.   0-486-44307-8

RADIOACTIVE SUBSTANCES, Marie Curie. The celebrated scientist's thesis, which directly preceded her 1903 Nobel Prize, discusses establishing atomic character of radioactivity; extraction from pitchblende of polonium and radium; isolation of pure radium chloride; more. 96pp. 5 3/8 x 8 1/2.   0-486-42550-9

CHEMICAL MAGIC, Leonard A. Ford. Classic guide provides intriguing entertainment while elucidating sound scientific principles, with more than 100 unusual stunts: cold fire, dust explosions, a nylon rope trick, a disappearing beaker, much more. 128pp. 5 3/8 x 8 1/2.   0-486-67628-5

ALCHEMY, E. J. Holmyard. Classic study by noted authority covers 2,000 years of alchemical history: religious, mystical overtones; apparatus; signs, symbols, and secret terms; advent of scientific method, much more. Illustrated. 320pp. 5 3/8 x 8 1/2.
0-486-26298-7

CHEMICAL KINETICS AND REACTION DYNAMICS, Paul L. Houston. This text teaches the principles underlying modern chemical kinetics in a clear, direct fashion, using several examples to enhance basic understanding. Solutions to selected problems. 2001 edition. 352pp. 8 3/8 x 11.   0-486-45334-0

PROBLEMS AND SOLUTIONS IN QUANTUM CHEMISTRY AND PHYSICS, Charles S. Johnson and Lee G. Pedersen. Unusually varied problems, with detailed solutions, cover of quantum mechanics, wave mechanics, angular momentum, molecular spectroscopy, scattering theory, more. 280 problems, plus 139 supplementary exercises. 430pp. 6 1/2 x 9 1/4.   0-486-65236-X

ELEMENTS OF CHEMISTRY, Antoine Lavoisier. Monumental classic by the founder of modern chemistry features first explicit statement of law of conservation of matter in chemical change, and more. Facsimile reprint of original (1790) Kerr translation. 539pp. 5 3/8 x 8 1/2.   0-486-64624-6

MAGNETISM AND TRANSITION METAL COMPLEXES, F. E. Mabbs and D. J. Machin. A detailed view of the calculation methods involved in the magnetic properties of transition metal complexes, this volume offers sufficient background for original work in the field. 1973 edition. 240pp. 5 3/8 x 8 1/2.   0-486-46284-6

GENERAL CHEMISTRY, Linus Pauling. Revised third edition of classic first-year text by Nobel laureate. Atomic and molecular structure, quantum mechanics, statistical mechanics, thermodynamics correlated with descriptive chemistry. Problems. 992pp. 5 3/8 x 8 1/2.   0-486-65622-5

ELECTROLYTE SOLUTIONS: Second Revised Edition, R. A. Robinson and R. H. Stokes. Classic text deals primarily with measurement, interpretation of conductance, chemical potential, and diffusion in electrolyte solutions. Detailed theoretical interpretations, plus extensive tables of thermodynamic and transport properties. 1970 edition. 590pp. 5 3/8 x 8 1/2.   0-486-42225-9

Engineering

FUNDAMENTALS OF ASTRODYNAMICS, Roger R. Bate, Donald D. Mueller, and Jerry E. White. Teaching text developed by U.S. Air Force Academy develops the basic two-body and n-body equations of motion; orbit determination; classical orbital elements, coordinate transformations; differential correction; more. 1971 edition. 455pp. 5 3/8 x 8 1/2. 0-486-60061-0

INTRODUCTION TO CONTINUUM MECHANICS FOR ENGINEERS: Revised Edition, Ray M. Bowen. This self-contained text introduces classical continuum models within a modern framework. Its numerous exercises illustrate the governing principles, linearizations, and other approximations that constitute classical continuum models. 2007 edition. 320pp. 6 1/8 x 9 1/4. 0-486-47460-7

ENGINEERING MECHANICS FOR STRUCTURES, Louis L. Bucciarelli. This text explores the mechanics of solids and statics as well as the strength of materials and elasticity theory. Its many design exercises encourage creative initiative and systems thinking. 2009 edition. 320pp. 6 1/8 x 9 1/4. 0-486-46855-0

FEEDBACK CONTROL THEORY, John C. Doyle, Bruce A. Francis and Allen R. Tannenbaum. This excellent introduction to feedback control system design offers a theoretical approach that captures the essential issues and can be applied to a wide range of practical problems. 1992 edition. 224pp. 6 1/2 x 9 1/4. 0-486-46933-6

THE FORCES OF MATTER, Michael Faraday. These lectures by a famous inventor offer an easy-to-understand introduction to the interactions of the universe's physical forces. Six essays explore gravitation, cohesion, chemical affinity, heat, magnetism, and electricity. 1993 edition. 96pp. 5 3/8 x 8 1/2. 0-486-47482-8

DYNAMICS, Lawrence E. Goodman and William H. Warner. Beginning engineering text introduces calculus of vectors, particle motion, dynamics of particle systems and plane rigid bodies, technical applications in plane motions, and more. Exercises and answers in every chapter. 619pp. 5 3/8 x 8 1/2. 0-486-42006-X

ADAPTIVE FILTERING PREDICTION AND CONTROL, Graham C. Goodwin and Kwai Sang Sin. This unified survey focuses on linear discrete-time systems and explores natural extensions to nonlinear systems. It emphasizes discrete-time systems, summarizing theoretical and practical aspects of a large class of adaptive algorithms. 1984 edition. 560pp. 6 1/2 x 9 1/4. 0-486-46932-8

INDUCTANCE CALCULATIONS, Frederick W. Grover. This authoritative reference enables the design of virtually every type of inductor. It features a single simple formula for each type of inductor, together with tables containing essential numerical factors. 1946 edition. 304pp. 5 3/8 x 8 1/2. 0-486-47440-2

THERMODYNAMICS: Foundations and Applications, Elias P. Gyftopoulos and Gian Paolo Beretta. Designed by two MIT professors, this authoritative text discusses basic concepts and applications in detail, emphasizing generality, definitions, and logical consistency. More than 300 solved problems cover realistic energy systems and processes. 800pp. 6 1/8 x 9 1/4. 0-486-43932-1

THE FINITE ELEMENT METHOD: Linear Static and Dynamic Finite Element Analysis, Thomas J. R. Hughes. Text for students without in-depth mathematical training, this text includes a comprehensive presentation and analysis of algorithms of time-dependent phenomena plus beam, plate, and shell theories. Solution guide available upon request. 672pp. 6 1/2 x 9 1/4. 0-486-41181-8

HELICOPTER THEORY, Wayne Johnson. Monumental engineering text covers vertical flight, forward flight, performance, mathematics of rotating systems, rotary wing dynamics and aerodynamics, aeroelasticity, stability and control, stall, noise, and more. 189 illustrations. 1980 edition. 1089pp. 5 5/8 x 8 1/4.   0-486-68230-7

MATHEMATICAL HANDBOOK FOR SCIENTISTS AND ENGINEERS: Definitions, Theorems, and Formulas for Reference and Review, Granino A. Korn and Theresa M. Korn. Convenient access to information from every area of mathematics: Fourier transforms, Z transforms, linear and nonlinear programming, calculus of variations, random-process theory, special functions, combinatorial analysis, game theory, much more. 1152pp. 5 3/8 x 8 1/2.   0-486-41147-8

A HEAT TRANSFER TEXTBOOK: Fourth Edition, John H. Lienhard V and John H. Lienhard IV. This introduction to heat and mass transfer for engineering students features worked examples and end-of-chapter exercises. Worked examples and end-of-chapter exercises appear throughout the book, along with well-drawn, illuminating figures. 768pp. 7 x 9 1/4.   0-486-47931-5

BASIC ELECTRICITY, U.S. Bureau of Naval Personnel. Originally a training course; best nontechnical coverage. Topics include batteries, circuits, conductors, AC and DC, inductance and capacitance, generators, motors, transformers, amplifiers, etc. Many questions with answers. 349 illustrations. 1969 edition. 448pp. 6 1/2 x 9 1/4.

0-486-20973-3

BASIC ELECTRONICS, U.S. Bureau of Naval Personnel. Clear, well-illustrated introduction to electronic equipment covers numerous essential topics: electron tubes, semiconductors, electronic power supplies, tuned circuits, amplifiers, receivers, ranging and navigation systems, computers, antennas, more. 560 illustrations. 567pp. 6 1/2 x 9 1/4.   0-486-21076-6

BASIC WING AND AIRFOIL THEORY, Alan Pope. This self-contained treatment by a pioneer in the study of wind effects covers flow functions, airfoil construction and pressure distribution, finite and monoplane wings, and many other subjects. 1951 edition. 320pp. 5 3/8 x 8 1/2.   0-486-47188-8

SYNTHETIC FUELS, Ronald F. Probstein and R. Edwin Hicks. This unified presentation examines the methods and processes for converting coal, oil, shale, tar sands, and various forms of biomass into liquid, gaseous, and clean solid fuels. 1982 edition. 512pp. 6 1/8 x 9 1/4.   0-486-44977-7

THEORY OF ELASTIC STABILITY, Stephen P. Timoshenko and James M. Gere. Written by world-renowned authorities on mechanics, this classic ranges from theoretical explanations of 2- and 3-D stress and strain to practical applications such as torsion, bending, and thermal stress. 1961 edition. 560pp. 5 3/8 x 8 1/2.   0-486-47207-8

PRINCIPLES OF DIGITAL COMMUNICATION AND CODING, Andrew J. Viterbi and Jim K. Omura. This classic by two digital communications experts is geared toward students of communications theory and to designers of channels, links, terminals, modems, or networks used to transmit and receive digital messages. 1979 edition. 576pp. 6 1/8 x 9 1/4.   0-486-46901-8

LINEAR SYSTEM THEORY: The State Space Approach, Lotfi A. Zadeh and Charles A. Desoer. Written by two pioneers in the field, this exploration of the state space approach focuses on problems of stability and control, plus connections between this approach and classical techniques. 1963 edition. 656pp. 6 1/8 x 9 1/4.

0-486-46663-9

Mathematics–Bestsellers

HANDBOOK OF MATHEMATICAL FUNCTIONS: with Formulas, Graphs, and Mathematical Tables, Edited by Milton Abramowitz and Irene A. Stegun. A classic resource for working with special functions, standard trig, and exponential logarithmic definitions and extensions, it features 29 sets of tables, some to as high as 20 places. 1046pp. 8 x 10 1/2. 0-486-61272-4

ABSTRACT AND CONCRETE CATEGORIES: The Joy of Cats, Jiri Adamek, Horst Herrlich, and George E. Strecker. This up-to-date introductory treatment employs category theory to explore the theory of structures. Its unique approach stresses concrete categories and presents a systematic view of factorization structures. Numerous examples. 1990 edition, updated 2004. 528pp. 6 1/8 x 9 1/4. 0-486-46934-4

MATHEMATICS: Its Content, Methods and Meaning, A. D. Aleksandrov, A. N. Kolmogorov, and M. A. Lavrent'ev. Major survey offers comprehensive, coherent discussions of analytic geometry, algebra, differential equations, calculus of variations, functions of a complex variable, prime numbers, linear and non-Euclidean geometry, topology, functional analysis, more. 1963 edition. 1120pp. 5 3/8 x 8 1/2. 0-486-40916-3

INTRODUCTION TO VECTORS AND TENSORS: Second Edition--Two Volumes Bound as One, Ray M. Bowen and C.-C. Wang. Convenient single-volume compilation of two texts offers both introduction and in-depth survey. Geared toward engineering and science students rather than mathematicians, it focuses on physics and engineering applications. 1976 edition. 560pp. 6 1/2 x 9 1/4. 0-486-46914-X

AN INTRODUCTION TO ORTHOGONAL POLYNOMIALS, Theodore S. Chihara. Concise introduction covers general elementary theory, including the representation theorem and distribution functions, continued fractions and chain sequences, the recurrence formula, special functions, and some specific systems. 1978 edition. 272pp. 5 3/8 x 8 1/2. 0-486-47929-3

ADVANCED MATHEMATICS FOR ENGINEERS AND SCIENTISTS, Paul DuChateau. This primary text and supplemental reference focuses on linear algebra, calculus, and ordinary differential equations. Additional topics include partial differential equations and approximation methods. Includes solved problems. 1992 edition. 400pp. 7 1/2 x 9 1/4. 0-486-47930-7

PARTIAL DIFFERENTIAL EQUATIONS FOR SCIENTISTS AND ENGINEERS, Stanley J. Farlow. Practical text shows how to formulate and solve partial differential equations. Coverage of diffusion-type problems, hyperbolic-type problems, elliptic-type problems, numerical and approximate methods. Solution guide available upon request. 1982 edition. 414pp. 6 1/8 x 9 1/4. 0-486-67620-X

VARIATIONAL PRINCIPLES AND FREE-BOUNDARY PROBLEMS, Avner Friedman. Advanced graduate-level text examines variational methods in partial differential equations and illustrates their applications to free-boundary problems. Features detailed statements of standard theory of elliptic and parabolic operators. 1982 edition. 720pp. 6 1/8 x 9 1/4. 0-486-47853-X

LINEAR ANALYSIS AND REPRESENTATION THEORY, Steven A. Gaal. Unified treatment covers topics from the theory of operators and operator algebras on Hilbert spaces; integration and representation theory for topological groups; and the theory of Lie algebras, Lie groups, and transform groups. 1973 edition. 704pp. 6 1/8 x 9 1/4. 0-486-47851-3

Browse over 9,000 books at www.doverpublications.com

A SURVEY OF INDUSTRIAL MATHEMATICS, Charles R. MacCluer. Students learn how to solve problems they'll encounter in their professional lives with this concise single-volume treatment. It employs MATLAB and other strategies to explore typical industrial problems. 2000 edition. 384pp. 5 3/8 x 8 1/2. 0-486-47702-9

NUMBER SYSTEMS AND THE FOUNDATIONS OF ANALYSIS, Elliott Mendelson. Geared toward undergraduate and beginning graduate students, this study explores natural numbers, integers, rational numbers, real numbers, and complex numbers. Numerous exercises and appendixes supplement the text. 1973 edition. 368pp. 5 3/8 x 8 1/2. 0-486-45792-3

A FIRST LOOK AT NUMERICAL FUNCTIONAL ANALYSIS, W. W. Sawyer. Text by renowned educator shows how problems in numerical analysis lead to concepts of functional analysis. Topics include Banach and Hilbert spaces, contraction mappings, convergence, differentiation and integration, and Euclidean space. 1978 edition. 208pp. 5 3/8 x 8 1/2. 0-486-47882-3

FRACTALS, CHAOS, POWER LAWS: Minutes from an Infinite Paradise, Manfred Schroeder. A fascinating exploration of the connections between chaos theory, physics, biology, and mathematics, this book abounds in award-winning computer graphics, optical illusions, and games that clarify memorable insights into self-similarity. 1992 edition. 448pp. 6 1/8 x 9 1/4. 0-486-47204-3

SET THEORY AND THE CONTINUUM PROBLEM, Raymond M. Smullyan and Melvin Fitting. A lucid, elegant, and complete survey of set theory, this three-part treatment explores axiomatic set theory, the consistency of the continuum hypothesis, and forcing and independence results. 1996 edition. 336pp. 6 x 9. 0-486-47484-4

DYNAMICAL SYSTEMS, Shlomo Sternberg. A pioneer in the field of dynamical systems discusses one-dimensional dynamics, differential equations, random walks, iterated function systems, symbolic dynamics, and Markov chains. Supplementary materials include PowerPoint slides and MATLAB exercises. 2010 edition. 272pp. 6 1/8 x 9 1/4. 0-486-47705-3

ORDINARY DIFFERENTIAL EQUATIONS, Morris Tenenbaum and Harry Pollard. Skillfully organized introductory text examines origin of differential equations, then defines basic terms and outlines general solution of a differential equation. Explores integrating factors; dilution and accretion problems; Laplace Transforms; Newton's Interpolation Formulas, more. 818pp. 5 3/8 x 8 1/2. 0-486-64940-7

MATROID THEORY, D. J. A. Welsh. Text by a noted expert describes standard examples and investigation results, using elementary proofs to develop basic matroid properties before advancing to a more sophisticated treatment. Includes numerous exercises. 1976 edition. 448pp. 5 3/8 x 8 1/2. 0-486-47439-9

THE CONCEPT OF A RIEMANN SURFACE, Hermann Weyl. This classic on the general history of functions combines function theory and geometry, forming the basis of the modern approach to analysis, geometry, and topology. 1955 edition. 208pp. 5 3/8 x 8 1/2. 0-486-47004-0

THE LAPLACE TRANSFORM, David Vernon Widder. This volume focuses on the Laplace and Stieltjes transforms, offering a highly theoretical treatment. Topics include fundamental formulas, the moment problem, monotonic functions, and Tauberian theorems. 1941 edition. 416pp. 5 3/8 x 8 1/2. 0-486-47755-X

Browse over 9,000 books at www.doverpublications.com